Studies in Systems, Decision and Control

Volume 409

Series Editor

The series "Studies in Systems, Decision and Control" (SSDC) covers both new developments and advances, as well as the state of the art, in the various areas of broadly perceived systems, decision making and control–quickly, up to date and with a high quality. The intent is to cover the theory, applications, and perspectives on the state of the art and future developments relevant to systems, decision making, control, complex processes and related areas, as embedded in the fields of engineering, computer science, physics, economics, social and life sciences, as well as the paradigms and methodologies behind them. The series contains monographs, textbooks, lecture notes and edited volumes in systems, decision making and control spanning the areas of Cyber-Physical Systems, Autonomous Systems, Sensor Networks, Control Systems, Energy Systems, Automotive Systems, Biological Systems, Vehicular Networking and Connected Vehicles, Aerospace Systems, Automation, Manufacturing, Smart Grids, Nonlinear Systems, Power Systems, Robotics, Social Systems, Economic Systems and other. Of particular value to both the contributors and the readership are the short publication timeframe and the world-wide distribution and exposure which enable both a wide and rapid dissemination of research output.

Indexed by SCOPUS, DBLP, WTI Frankfurt eG, zbMATH, SCImago.

All books published in the series are submitted for consideration in Web of Science.

More information about this series at https://link.springer.com/bookseries/13304

Deyan Radev

Measuring Systemic Risk

A Probabilistic Perspective

Springer

Deyan Radev
Faculty of Economics and Business
Administration
Sofia University
Sofia, Bulgaria

ISSN 2198-4182 ISSN 2198-4190 (electronic)
Studies in Systems, Decision and Control
ISBN 978-3-030-94280-9 ISBN 978-3-030-94281-6 (eBook)
https://doi.org/10.1007/978-3-030-94281-6

This Springer imprint is published by the registered company Springer Nature Switzerland AG
The registered company address is: Gewerbestrasse 11, 6330 Cham, Switzerland

Acknowledgements

My research on probability-derived systemic risk measures that is the basis for this book was awarded in 2014 with the Best Dissertation Award of Deutsche Bundesbank and the Hochschulpreis in Finance of Deutsches Aktieninstitut. The systemic risk measures in the following pages are in daily operational use by the European Central Bank (ECB) and the European Systemic Risk Board (ESRB). I would like to thank these institutions for the acknowledgment and support for my research.

I would like to thank Isabel Schnabel for her support and mentorship. The research underlying this book was partially conducted while I was a guest researcher at the Financial Stability Surveillance Division of the European Central Bank. I would like to thank Puriya Abbassi, Toni Ahnert, Pooyan Amir Ahmadi, Ivan Alves, Bernd Bartels, Andreas Barth, Frank Betz, Michael Binder, Carsten Detken, Paola Donati, Andrea Gamba, Paolo Giudici, Denis Gorea, Charles Goodhart, Martin Götz, Reint Gropp, Paul Hiebert, Jan-Pieter Krahnen, Melanie Krause, Silviu Oprica, Laura Parisi, Loriana Pelizzon, Rafael Repullo, David Schumacher, Ctirad Slawik, Constantin Weiser, Mirko Wiederholt and Chen Zhou for their valuable discussions and recommendations. This book also benefited from the suggestions and comments of the ECB FSS staff and the participants of the 1st Workshop in Financial Economics, the Brown Bag seminar at Johannes Gutenberg University Mainz, the 2012 GSEFM Summer Institute, the 2012 Conferences of the German Economic Association and the German Finance Association, the 2013 Congress of the European Economic Association, the Eurofidai December 2013 Paris Finance Meeting, the 2014 IFABS Lisbon conference, the 2016 ESRB and Bank of Finland Conference in Helsinki and the 5th Emerging Scholars in Banking and Finance Conference at the Business School, London (formerly Cass Business School). All remaining errors are my own. I gratefully acknowledge research support from the Leibniz Institute for Financial Research SAFE, funded by the State of Hessen initiative for research LOEWE.

Contents

List of Figures

List of Tables

Chapter 1
Introduction

The issue of sovereign default has grown in importance in the past decade, especially in the euro area (EA). The fear of the potential negative consequences of a default of an EA government has led to a number of sovereign and bank bailouts and affected interest rates, capital flows, trade, and economic growth in Europe and around the globe. The necessity for an in-depth analysis of the level of sovereign risk and how it affects the broader financial system in the EA has raised a number of fundamental questions: How to quantify systemic risk of sovereigns? How to measure the feedback and cascade effects from a sovereign default? Would defaults of small periphery governments be followed by a default of a large EA sovereign? What are the main drivers of systemic risk in the euro area? Despite being of utmost importance for consistent and timely macro- and microprudential policies, these questions remain unanswered. This book contributes to this discussion with an innovative approach to analyze systemic risk and contagion among sovereigns and banks, based on market expectations about default risk. Using that approach, we develop a number of important systemic risk measures for the regulators' toolkit.

An attractive approach to modeling default risk is to view the financial system as a joint distribution of its constituting financial institutions, as it allows publicly available market data to be incorporated in comprehensive procedures to assess the likelihood of extreme systemic events. Zhou (2010) stresses the inability of most systemic risk measures, such as the Conditional Value-at-Risk (Adrian and Brunnermeier 2016), the Marginal Expected Shortfall (Acharya et al. 2009, 2012) and the SRISK (Brownlees and Engle 2017) to account for multivariate interactions, as they focus on bilateral relationships either between two financial institutions or between an aggregated system index and an individual institution. A second major drawback of most measures is that they focus on measuring the risk of financial institutions, ignoring the systemic risk of sovereigns. Another major advantage of our approach to the CoVaR and its related measures is that we use information regarding market expectations on default, while the CoVaR uses historical stock market data. Giglio (2016) points out that reduced-form approaches, recovering return distributions from

D. Radev, *Measuring Systemic Risk*, Studies in Systems, Decision and Control 409,
https://doi.org/10.1007/978-3-030-94281-6_1

historical data, as the CoVaR, suffer from the low number of extreme events in market data. In contrast, our approach circumvents this issue, by recovering default probabilities from derivatives which are more sensitive to default risk, such as CDS or option contracts.

The procedure we propose consists of three steps. First, we use CDS spreads to extract the perceived individual default risk of 10 euro area (EA) sovereigns and 15 EA large and complex banking institutions (LCBGs) using a CDS bootstrapping procedure (Hull and White 2000). In a second step, we model the multivariate probability densities of the EA banking and sovereign systems in a way that is consistent with the individual probabilities of default (PoDs). Among the various procedures discussed in the literature, the method that we consider especially appropriate for our purposes is the minimum cross-entropy procedure developed by Kullback (1959) and extended by Segoviano (2006) and Segoviano and Goodhart (2009).[1] This methodology has solid conceptual underpinnings, allowing us to focus on the market beliefs of the performance of a bank or a sovereign, while avoiding a direct examination of their capital structure. With respect to sovereign default risk estimation, this appears to be a more attractive alternative to the Sovereign Contingent Claims Analysis of Gray et al. (2007) and Gray (2011), where the authors try to sort the capital structure of a sovereign in a particular way depending on its maturity, in order to fit it in a Merton Model framework (Merton 1974). This requires the assets and liabilities to be assigned to a category at every given point in time, making the method relatively cumbersome. By applying the cross-entropy method, we avoid this procedure by focusing directly on probabilities of default derived from market data and assuming a standardized distribution for our initial beliefs about the individual entity's assets. However, we still rely on the intuition of the Merton Model, that an entity (a bank or a sovereign) defaults on its debt, once its assets can no longer cover its liabilities. After recovering the system's multivariate probability density, we proceed to our final stage—using variations of the Bayes' rule to derive a series of systemic risk indicators that analyze the fragility of the financial system to default events.

We should point out that the calculation of the probability measures is independent from the estimation procedure of choice. Instead of our minimum cross-entropy approach, it can be calculated, for instance, by using a multivariate t-distribution as in Lucas et al. (2014, 2017), and such a method will employ variable thresholds, as opposed to the fixed ones in the minimum cross-entropy approach. Unreported robustness checks showed minor quantitative differences between both calculation methods, while the qualitative conclusions remained the same.

We rely on CDS data, because, first, bank and sovereign defaults are relatively rare events and it is difficult to arrive at meaningful actual default frequencies even for entities at the brink of insolvency. The problem can be circumvented by using CDS-derived expected probabilities of default. Our bootstrapping procedure allows us to arrive at expected probability of default using a single CDS spread observation, which means that we reduce the input data requirements significantly, compared to

[1] Segoviano (2006) calls the method Consistent Information Multivariate Density Optimization (CIMDO).

the CoVaR, the MES and the SRISK, which rely on large amounts of data to estimate historical default frequencies and risk exposures. This brings us to our second motive for using bank and sovereign CDS spreads: we analyze *market expectations* about *individual* default risk and to develop a model that transforms them into *joint* bank and/or sovereign default risk perceptions. Third, due to the numerous interventions on the EA bond market in recent times, bond prices are not suitable for analyzing expectations about individual or joint default risk. No interventions are undertaken or planned on the CDS market, so the CDS premia are the true market consensus prices of default risk.

It is important to note that by using CDS-derived default probabilities, we are concentrating on market default risk *perceptions* and not directly on *actual* default frequencies. Market perceptions about individual and joint default risk are of interest to investors and regulators, because default risk premia comprise a large part of bond yields. Since nowadays banks and sovereigns rely heavily on international financial markets to finance their liquidity needs, they should pay closer attention to market expectations about credit risk. The recent developments on the CDS and bond markets show that investors do not only price in individual banking and sovereign characteristics, but also take into account the interdependence among these actors. This is especially important for regulators in their analysis of the effectiveness of policy measures. Furthermore, since the PoDs extracted from CDS spreads are risk-neutral and thus overestimate actual (physical) probabilities, we arrive at relatively conservative default measures. This is of particular importance for macroprudential policies, where regulators need to be alerted in advance for possible vulnerabilities in the financial system.

We contribute to the existing literature in a number of ways. First, we add to a broader research agenda with the main purpose to shed more light on the distress vulnerability of the EA financial system during the recent crises. Gorea and Radev (2014) and Acharya and Steffen (2015) investigate the determinants of the default risk of euro area countries and find that investors discriminate between the EA "core" (Austria, Belgium, France, Germany and Netherlands) and "periphery" (Greece, Ireland, Italy, Portugal and Spain) already during the Post-Lehman global recession, and almost two years before the Sovereign debt crisis. Second, the current work complements and extends the regulatory toolbox of systemic risk measures, such as the CoVaR, the Marginal Expected Shortfall (Acharya et al. 2009, 2012) and the SRISK (Brownlees and Engle 2017). In particular, we add to the multivariate probability measures introduced by Avesani et al. (2006), Segoviano and Goodhart (2009), Giglio (2016), Lucas et al. (2014) and Radev (2019). We believe that our study can help policy makers and regulators to get a more comprehensive perspective on the EA systemic risk vulnerabilities in the current turbulent times.

Third, we introduce a new and intuitive systemic risk measure that evaluates the contribution of the system's interdependence to systemic default risk. Similar to the Shapley value, an important feature of our measure is the flexibility in the choice of the "coalitions" of defaulting and non-defaulting sovereigns. This means that we could develop a large number of $\Delta CoJPoD$ measures for a single sample of sovereigns or companies, conditioning on the default of one, two or more governments or firms. Therefore, $\Delta CoJPoD$ could be seen as a *family* of measures of systemic risk, rather than as a single measure. We illustrate this flexibility we investigate cascade effects among euro area sovereigns in Sect. 5.3. Fourth, we contribute to the theory on maximizing cross-entropy and extend it to sovereign default. Fifth, we extend the relatively sparse literature on sovereign default in a multivariate setting by investigating not only the contribution of individual sovereigns to the systemic default risk, but also possible cascade effects within the euro area.

The remainder of this book is organized as follows.

Chapter 2 Provides a discussion of the literature on systemic risk and the previous work that is most related to our approach.

Chapter 3 describes in detail our three-step procedure to derive systemic risk measures based on probabilities of default. First, we show how we derive individual PoDs from CDS spreads using a bootstrapping procedure (Hull and White 2000). Second, we outline how use a minimum cross-entropy approach to construct a multivariate probability distribution consistent with the individual probabilities of default (Kullback 1959; Segoviano 2006). Once we have recovered the multivariate density, in the following chapters, we introduce a number of systemic distress measures that trace narrowly the feedback effects within the euro area's system of sovereigns and banks.

Chapter 4 introduces several unconditional and conditional measures of joint default risk in the sovereign and banking systems. The measures vary in their level of complexity and interpretation: From probabilities of default of a sovereign (bank), conditional on one or two sovereigns (banks) defaulting, to probabilities of at least $N - 1$ sovereigns (banks) defaulting, given a sovereign (bank) defaults. The chapter also introduces the Systemic Fragility Measure: The probability of at least two sovereigns or banks defaulting. Our approach to calculate the measure made it possible to incorporate the indicator in the ECB's and ESRB's toolkit and it is in daily operational use by the Bank and the Board since February 2012. We provide a sensitivity analysis of that particular measure to changes in a number of important assumptions about the dependence structure among sovereigns or banks and on the shape of the joint distribution of assets in the financial system.

In **Chap.** 5, we devise a new and intuitive measure of systemic risk *contributions* based on the *information content* in a default of a sovereign in an interdependent financial system. The measure can be viewed as a bias-corrected perception of the dynamics of systemic risk after a default of a sovereign. We apply it to estimate the effect of individual and joint sovereign default on the overall default risk in the euro area system of sovereigns. We find that sovereign contributions to default increase after Lehman and especially during the Sovereign debt crisis, with a considerable potential for cascade effects among euro area sovereigns. The new measure can help policymakers to assess the impact of macro- and microprudential policies. It has a wide range of applications, such as the investigation of spillover effects from sovereigns to financial institutions, or among industries in a country or region.

Chapter 6 wraps up the analysis, analyzes the state of development of systemic risk measurement and outlines a number of venues for future research and policy applications.

References

Acharya, V.V., Steffen, S.: The greatest carry trade ever? Understanding eurozone bank risks. J. Financ. Econ. **115**(2), 215–236 (2015)

Acharya, V., Pedersen, L., Philippon, T., Richardson, M.: Regulating systemic risk. In: Acharya, V., Richardson, M. (eds.) Restoring Financial Stability: how to Repair a Failed System. Wiley, New York (2009)

Acharya, V., Pedersen, L., Philippon, T., Richardson, M.: Measuring systemic risk. CEPR Discussion Papers 8824. C.E.P.R. Discussion Papers (2012)

Adrian, T., Brunnermeier, M.: CoVaR. Am. Econ. Rev. **106**(7), 1705–1741 (2016)

Avesani, R., Li, J., Pascual, A.: A new risk indicator and stress testing tool: a multifactor Nth-to-default CDS basket. IMF Working Papers 06/105. International Monetary Fund (2006)

Brownlees, C., Engle, R.F.: SRISK: a conditional capital shortfall measure of systemic risk. Rev. Financ. Stud. **30**(1), 48–79 (2017)

Giglio, S.: Credit default swap spreads and systemic financial risk. ESRB Working Paper Series 15. European Systemic Risk Board (2016)

Gorea, D., Radev, D.: The euro area sovereign debt crisis: can contagion spread from the periphery to the core? Int. Rev. Econ. Financ. **30**, 78–100 (2014)

Gray, D.: Enhanced bank stress testing incorporating sovereign risk, funding cost, and liquidity risk using contingent claims analysis. Background Paper. ESRB Advisory Technical Committee Workshop (2011)

Gray, D., Bodie, Z., Merton, R.: Contingent claims approach to measuring and managing sovereign risk. J. Invest. Manag. **5**(4), 5–28 (2007)

Hull, J., White, A.: Valuing credit default swaps I: no counterparty default risk. J. Deriv. **8**(1), 29–40 (2000)

Kullback, S.: Information Theory and Statistics. Wiley, Chichester (1959)

Lucas, A., Schwaab, B., Zhang, X.: Conditional probabilities and contagion measures for euro area sovereign default risk. J. Bus. Econ. Stat. **32**, 271–284 (2014)

Lucas, A., Schwaab, B., Zhang, X.: Modeling financial sector joint tail risk in the euro area. J. Appl. Econom. **32**, 171–191 (2017)

Merton, R.: On the pricing of corporate debt: the risk structure of interest rates. J. Financ. **29**(2), 449–470 (1974)

Radev, D.: Systemic risk, sovereign default and cascade effects in the euro area. Technical report. University of Bonn (2019)

Segoviano, M.: Consistent information multivariate density optimizing methodology. FMG Discussion Papers 557. International Monetary Fund (2006)

Segoviano, M., Goodhart, C.: Banking stability measures. IMF Working Paper 09/4. International Monetary Fund (2009)

Zhou, C.: Are banks too big to fail? Measuring systemic importance of financial institutions. Int. J. Cent. Bank. **6**(34), 205–250 (2010)

Chapter 2
Related Literature

The literature on systemic risk focuses primarily on measuring the systemic consequences of a default of *financial intermediaries*. The most widely accepted measures of systemic risk are the CoVaR (Adrian and Brunnermeier 2016), the Marginal Expected Shortfall (Acharya et al. 2009, 2012), and the SRISK (Brownlees and Engle 2017). While being highly effective in gauging different dimensions of systemic risk stemming from distressed financial companies, these indicators are based on stock market and balance sheet data, which are rarely available for sovereigns. Therefore, the traditional systemic risk measures cannot be directly applied to quantify sovereign default risk. In this book, we solve this problem by deriving probabilities of default (PoDs) from credit default swap (CDS) data that is readily available for sovereigns.

Our book is related to several strands of literature. First, our approach is closely related to the literature on measuring contagion and systemic risk using entropy maximization within networks of banks and sovereigns. Using a unique dataset of bilateral exposures, Mistrulli (2011) applies the method to analyze how contagion propagates within the Italian interbank market. The author finds that interbank linkages contribute to contagion effects and that the performance of the maximum entropy approach depends on the size of the market, the presence of large banks and the loss rate. Castren and Rancan (2014) apply maximum entropy to macro-network analysis and find that the contagion effects depend on the underlying network structure, which changes over time. The authors identify sectors that are more conductive to shock propagation and argue that a more diversified portfolio of cross-border holdings may mitigate contagion. Paltalidis et al. (2015) extend the models by Mistrulli (2011) and Castren and Rancan (2014) by analyzing the losses and cascade effects stemming from sovereign credit risk and asset price risk. While the authors of these previous studies manage to model complex interactions and propagation effects within the financial system, their analysis is limited by the low frequency of the data that

D. Radev, *Measuring Systemic Risk*, Studies in Systems, Decision and Control 409,
https://doi.org/10.1007/978-3-030-94281-6_2

they use.[1] Our CDS-based approach allows us to derive probabilities of default and dynamic interactions at a daily frequency. The conditional joint probabilities can then be coupled with low-frequency data to calculate fairly accurate daily exposures. This extension is beyond the scope of the current book.

The second strand of the literature that our book relates to is the literature on estimating joint probabilities of default in the financial system. Segoviano and Goodhart (2009) and Radev (2014) develop a number of measures to analyze the joint default risk of banks using CDS data. Gray et al. (2007) and Gray (2011) develop the so called Sovereign Contingent Claims Analysis where the authors try to sort the capital structure of a sovereign in a particular way depending on its maturity, in order to fit it to a Merton model's framework (Merton 1974). This requires the assets and liabilities to be assigned to a category at every given point in time, making the method relatively cumbersome. In the current book, we argue that relying on market perceptions about default, embedded in CDS premia, appears to be an attractive alternative to the Sovereign Contingent Claims Analysis, because we avoid this sorting procedure. Notwithstanding, we still rely on the intuition of the Merton model that an entity (in our case—a sovereign) defaults on its debt, once its assets process crosses a certain default threshold.

Third, in developing $\Delta CoJPoD$ particularly, we draw from the theory on information contagion in the banking literature, developed by Acharya and Yorulmazer (2008). This theory postulates that a poor performance of a particular bank may contain valuable information about the performance of the other banks in the system and that healthy banks can also be negatively affected if they are considered too similar to the ailing bank (Acharya and Yorulmazer 2008; Ahnert and Georg 2017). The connection to $\Delta CoJPoD$ is based on the observation that the conditional probability of default has two ingredients—a potentially informative and an uninformative part. The informative portion provides additional information about the vulnerability of the euro area system, conditional on the default of a sovereign, and it needs to be disentangled from the uninformative portion, where the default of the sovereign has no impact on the system. The $\Delta CoJPoD$ focuses on the informative portion of conditional default probabilities and therefore represents the *information content* in a sovereign default. In Sect. 5.5.1 we show that focusing solely on conditional probabilities without a consideration about the actual additional information that they provide could lead to *spurious conclusions* and a bias in the estimation of the actual systemic risk. This feature of conditional default probabilities has not yet been analyzed in the financial stability literature.

Finally, $\Delta CoJPoD$ is conceptually related to other measures of systemic risk like the CoVaR (Adrian and Brunnermeier 2016), the Shapley value (Tarashev et al. 2010), the Marginal Expected Shortfall (Acharya et al. 2009, 2012), and the SRISK (Brownlees and Engle 2017), which view systemic risk contributions as the difference in the *value at risk* (VaR) of the system when an entity defaults, compared to the case when no default occurs. The main difference to these three concepts is that while they

[1] The highest frequency of the data in Mistrulli (2011), Castren and Rancan (2014) and Paltalidis et al. (2015) is monthly.

focus on conditional *value at risk* (the CoVaR) and conditional *expected shortfall* (the Shapley value, the Marginal Expected Shortfall and SRISK), the objects of our analysis are conditional *probabilities of default*. In concentrating on probabilities of default, our approach has several important advantages over the CoVaR and its related measures. First, the previous measures rely on a restrictive definition of default: an institution is considered under distress if its returns drop to the 5-percent or the 1-percent region of its return distribution. In contrast, our approach does not take a stand on what a default actually means, but rather relies on market expectations about the likelihood of a default to occur, based on an unobservable latent process. The main benefit of such an approach is that it allows us to derive systemic risk measures in a *sovereign* context without explicitly defining what sovereign assets are. Since systemic risk is usually defined as the risk of the collapse of the financial system if one (or more) of its participants defaults, focusing on probabilities of default allows us to address the definition more directly, and, in addition, to avoid the problem with the measurement of sovereign assets.

Another drawback of the CoVaR, the Shapley value, the MES and SRISK is that they rely on historical stock market data. Giglio (2016) points out that reduced-form approaches, recovering return distributions from historical data, suffer from the low number of extreme events in market data. Even if such events existed for euro area sovereigns, their low frequency would at best yield static estimates of probabilities of default. Hence, we would not be able to capture the changes in systemic risk over time, and especially after regulatory interventions. In contrast, our approach circumvents this issue by recovering forward-looking default probabilities from derivatives which are more sensitive to default risk, such as CDS contracts. The additional benefit of probabilities of default derived from CDS data is that they also reflect market expectations about sovereign default. Monitoring financial market expectations is of significant importance for policy decision-making in the euro area, since the governments in the eurozone rely on financial markets to finance their short and long term liquidity needs. Moreover, the reactions of the financial markets are an important indicator of how viable and credible policy measures are.

References

Acharya, V., Yorulmazer, T.: Information contagion and bank herding. J. Money Credit Bank. **40**, 215–231 (2008)

Acharya, V., Pedersen, L., Philippon, T., Richardson, M.: Regulating systemic risk. In: Acharya, V., Richardson, M. (eds.) Restoring Financial Stability: how to Repair a Failed System. Wiley, New York (2009)

Acharya, V., Pedersen, L., Philippon, T., Richardson, M.: Measuring systemic risk. CEPR Discussion Papers 8824. C.E.P.R. Discussion Papers (2012)

Adrian, T., Brunnermeier, M.: CoVaR. Am. Econ. Rev. **106**(7), 1705–1741 (2016)

Ahnert, T., Georg, C.-P.: Information contagion and systemic risk. J. Financ. Stab. forthcoming (2017)

Brownlees, C., Engle, R.F.: SRISK: a conditional capital shortfall measure of systemic risk. Rev. Financ. Stud. **30**(1), 48–79 (2017)

Castren, O., Rancan, M.: Macro-networks: an application to euro area financial accounts. J. Bank. Financ. **46**(C), 43–58 (2014)

Giglio, S.: Credit default swap spreads and systemic financial risk. ESRB Working Paper Series 15. European Systemic Risk Board (2016)

Gray, D.: Enhanced bank stress testing incorporating sovereign risk, funding cost, and liquidity risk using contingent claims analysis. Background Paper. ESRB Advisory Technical Committee Workshop (2011)

Gray, D., Bodie, Z., Merton, R.: Contingent claims approach to measuring and managing sovereign risk. J. Invest. Manag. **5**(4), 5–28 (2007)

Merton, R.: On the pricing of corporate debt: the risk structure of interest rates. J. Financ. **29**(2), 449–470 (1974)

Mistrulli, P.E.: Assessing financial contagion in the interbank market: maximum entropy versus observed interbank lending patterns. J. Bank. Financ. **35**(5), 1114–1127 (2011)

Paltalidis, N., Gounopoulos, D., Kizys, R., Koutelidakis, Y.: Transmission channels of systemic risk and contagion in the European financial network. J. Bank. Financ. **61**(S1), 36–52 (2015)

Radev, D.: Assessing systemic fragility: a probabilistic perspective. SAFE Working Paper 70. Research Center SAFE, Goethe University Frankfurt (2014)

Segoviano, M., Goodhart, C.: Banking stability measures. IMF Working Paper 09/4. International Monetary Fund (2009)

Tarashev, N., Borio, C., Tsatsaronis, K.: Attributing systemic risk to individual institutions. BIS Working Papers 308. Bank for International Settlements (2010)

Chapter 3
Multivariate Probabilities from Individual CDS Spreads

Our procedure to estimate multivariate probability measures involves three steps. First, we derive individual PoDs from CDS spreads using a bootstrapping procedure (Hull and White 2000). Second, we use a minimum cross-entropy approach to construct a multivariate probability distribution consistent with the individual probabilities of default (Kullback 1959; Segoviano 2006). Once we have recovered the multivariate density, in the following chapters we introduce a number of systemic distress measures, through which we can trace narrowly the feedback effects within the euro area's system of sovereigns and banks.

3.1 CDS Bootstrapping Procedure

In this exposition, we use a *refined* way of estimating probabilities of default (PoD), the CDS bootstrapping.[1] The procedure follows Hull and White (2000) and is based on a simple cumulative probability model, which incorporates recovery rates,

[1] The usual method for estimating probabilities of default from CDS spreads used in the literature is to use the most liquid contracts on the market, 5-year CDS spreads, to estimate one-year probabilities of default, applying the simple formula

$$PoD_t = \frac{CDS_t * 0.0001}{1 - RecoveryRate}, \tag{3.1}$$

where CDS_t is the 5-year CDS spread at time t, PoD_t is the resulting probability of default estimate and $RecoveryRate$ is an assumed recovery rate of the face value of the underlying bond in case of default. As only the first of the five annual premia is used in the formula, it is believed that the resulting series reflect accurately the one-year probabilities of default.

D. Radev, *Measuring Systemic Risk*, Studies in Systems, Decision and Control 409,
https://doi.org/10.1007/978-3-030-94281-6_3

refinancing rates and cumulative compounding. The method uses CDS contracts of different maturities to calibrate hazard rates of particular time horizons to estimate cumulative probabilities of default.[2] This method could be used for both sovereign and corporate probability of default estimation. The resulting risk measures are risk-neutral probabilities of default and satisfy the no-arbitrage condition in financial markets.

We follow Hull and White (2000) and derive our marginal probabilities of default using an iterative procedure called CDS bootstrapping. This section is intended to convey the main intuition behind the bootstrapping procedure. For a more detailed outline with formal derivations, please refer to the original article.

If τ is the time to default, one can express the probability of default at market date t, PoD(t) as

$$PoD(t) = P\left[\tau \leq t\right] = 1 - P\left[\tau \geq t\right] = 1 - Q(t), \tag{3.2}$$

where $P\left[\tau \leq t\right]$ and $P\left[\tau \geq t\right] = Q(t)$ are the probability of the time to default to be less than t and the survival probability, respectively.

We apply a standard survival probability model by expressing $Q(t)$ via a piecewise constant hazard rate $h(t)$. For instance, given that

$$h(t) = \begin{cases} \gamma_1 & \text{for } 0 \leq t \leq t_1, \\ \gamma_2 & \text{for } t_1 < t \leq t_2, \\ \gamma_3 & \text{for } t_2 < t, \end{cases} \tag{3.3}$$

the survival function is

$$Q(t) = \begin{cases} e^{\gamma_1 \cdot t} & \text{for } 0 \leq t \leq t_1, \\ e^{\gamma_1 \cdot t - \gamma_2 \cdot (t - t_1)} & \text{for } t_1 < t \leq t_2, \\ e^{\gamma_1 \cdot t_1 - \gamma_2 \cdot (t_2 - t_1) - \gamma_3 \cdot (t - t_2)} & \text{for } t_2 < t, \end{cases} \tag{3.4}$$

The CDS bootstrapping procedure then calibrates γ_1, γ_2 and γ_3 to the market CDS premia S_1, S_2 and S_3, such that the present value of the payment in case of default (100%—recovery rate), called also the "protection leg", equals the discounted premia flows in the CDS contract, or the "premium leg" at the given market dates t_1, t_2 and t_3. This equality relies on the no-arbitrage condition on financial markets. In practice this is an iterative procedure that starts with the shortest maturity contract to calculate the first hazard rate, γ_1, and works its way through to the longest maturity, making sure that the no-arbitrage condition holds at each step. In our calculation, we also account for the quarterly structure of the CDS contract and the accrued premium that should be paid, given the default is anywhere in between any two market dates. We then use Eq. (3.2) to calculate our cumulative $PoD(t)$, with $t = 1, \ldots, T$ denoting the default *horizon* (T = 5 years in our case). Effectively, we use $PoD(T)$ and annualize

[2] We outline CDS bootstrapping in more detail in Sect. 3.1.

it using Eq. (4.1) (note the innocuous abuse of notation: in the text, we use index t to denote each date in our sample, while here, we use it to denote a particular time horizon).

3.2 Recovery of the Multivariate Probability Density

Since joint credit events are rarely traded in the default insurance market, we need to impose a certain structure on the system's joint probability density, to allow the transition from individual to joint probabilities of default.

We base our methodology on the concept of cross-entropy, introduced by Kullback (1959). Our method minimizes a cross-entropy objective function that updates an agnostic prior distribution to arrive at a posterior distribution governed by a set of constraints. The constraints are chosen to ensure that the posterior distribution is consistent with the individual PoDs. This technique allows us to adjust our prior guess of the joint distribution function. We start by assuming a joint normal density function for our prior,[3] with a correlation matrix estimated from market data. We arrive at a posterior density that mirrors the behavior of the prior in particular areas of the distribution, but still reflects the possibly higher probability mass in the default region. The latter property reflects the well-documented fact that financial markets are characterized by a higher number of crashes than predicted by the normal distribution. The resulting *posterior* joint distribution has two main properties: first, it reflects the market consensus views about the default region of the unobserved asset distribution of the system, and second, it possesses fat tails, even if our starting assumption is a joint normal distribution. The cross-entropy approach is consistent with the individual PoDs data regardless of the assumed prior.

Segoviano (2006), Gorea and Radev (2014) and Radev (2014) analyze the robustness of the minimum cross-entropy approach with respect to some of its main underlying parameters, namely the prior distribution and dependence structure assumptions, paying special attention to the performance in the default region of the posterior joint distribution. Their results show that assuming a multivariate standard normal distribution as a prior provides very similar results to employing a fatter-tail distribution. The resulting posterior distribution is sufficient to explain the behavior in the default region of the distribution of sovereign assets. Assuming a more complex prior distribution does not provide a significant improvement. Therefore, we decide to use a multivariate joint normal distribution for our prior.

[3] In Sect. 4.5, we show that the results are not affected if we choose a fat-tailed distribution as a prior.

Let the financial system be represented by a portfolio[4] of n entities (sovereigns or banks): X_1, X_2 to X_n, with their log-assets being x_1, x_2, to x_n. The cross-entropy approach then minimizes the following Lagrangian:

$$
\begin{aligned}
L(p,q) = & \int_{-\infty}^{+\infty}\int_{-\infty}^{+\infty}\cdots\int_{-\infty}^{+\infty} p(x_1,x_2,\ldots,x_n)\ln\left[\frac{p(x_1,x_2,\ldots,x_n)}{q(x_1,x_2,\ldots,x_n)}\right]dx_1\ldots dx_{n-1}dx_n \\
& + \lambda_1\left[\int_{-\infty}^{+\infty}\int_{-\infty}^{+\infty}\cdots\int_{-\infty}^{+\infty} p(x_1,x_2,\ldots,x_n)\mathbf{I}_{[\overline{\mathbf{x}}_1,\infty)}dx_1\ldots dx_{n-1}dx_n - PoD_t^1\right] \\
& + \lambda_2\left[\int_{-\infty}^{+\infty}\int_{-\infty}^{+\infty}\cdots\int_{-\infty}^{+\infty} p(x_1,x_2,\ldots,x_n)\mathbf{I}_{[\overline{\mathbf{x}}_2,\infty)}dx_1\ldots dx_{n-1}dx_n - PoD_t^2\right] \\
& + \cdots \\
& + \lambda_n\left[\int_{-\infty}^{+\infty}\int_{-\infty}^{+\infty}\cdots\int_{-\infty}^{+\infty} p(x_1,x_2,\ldots,x_n)\mathbf{I}_{[\overline{\mathbf{x}}_n,\infty)}dx_1\ldots dx_{n-1}dx_n - PoD_t^n\right] \\
& + \mu\left[\int_{-\infty}^{+\infty}\int_{-\infty}^{+\infty}\cdots\int_{-\infty}^{+\infty} p(x_1,x_2,\ldots,x_n)dx_1\ldots dx_{n-1}dx_n - 1\right]
\end{aligned}
\qquad (3.5)
$$

The first integral in Eq. (3.5) represents the cross-entropy probability difference, introduced in Kullback (1959). The idea of this approach is to minimize the distance between a *prior* distribution guess $q(x_1, x_2, \ldots, x_n) \in \mathbb{R}^n$ and a *posterior* distribution $p(x_1, x_2, \ldots, x_n) \in \mathbb{R}^n$ that reflects empirical market data on individual probabilities of default. The market information is included in the model by means of consistency constraints (the lines from the second to the last but one in Eq. (3.5)). In those constraints, PoD_t^1, PoD_t^2 to PoD_t^n stand for the expected probabilities of default of the respective entities, derived from CDS prices. With $\mathbf{I}_{[\overline{\mathbf{x}}_1,\infty)}$, $\mathbf{I}_{[\overline{\mathbf{x}}_2,\infty)}$ to $\mathbf{I}_{[\overline{\mathbf{x}}_n,\infty)}$ we denote a set of indicator functions that take the value of one if the respective entities' default thresholds $\overline{\mathbf{x}}_1$, $\overline{\mathbf{x}}_2$ to $\overline{\mathbf{x}}_n$ are crossed and zero in the opposite case. The empirical default thresholds are analogous to the ones in the classic structural model (Merton 1974) and are calculating by inverting a standard normal cumulative distribution at the borrower's average estimated PoD for the sample period. To constrain the posterior distribution to the $[0,\ 1]^n$ region, we need to impose the additivity rule in the last line of Eq. (3.5). The coefficients μ, λ_1, λ_2 to λ_n are the Lagrange multipliers of the described constraints. The optimal posterior distribution is then represented in the following way[5]:

[4] The current section outlines the multivariate case of the cross-entropy approach. For lower dimensionality problems, see Segoviano (2006) and Gorea and Radev (2014).

[5] See Sect. 3.2.1 for a complete solution of the multivariate minimum cross-entropy problem.

$$p^*(x_1, x_2, \ldots, x_n) = q(x_1, x_2, \ldots, x_n) exp\left\{-\left[1 + \mu + \sum_{i=1}^{n} \lambda_i \mathbf{I}_{[\bar{\mathbf{x}}_i, \infty)}\right]\right\} \tag{3.6}$$

Hence, in order to derive the optimal posterior distribution, all we need is the prior distribution with an appropriate dependence structure (multivariate Gaussian density with an empirical correlation matrix), the optimal Lagrange multipliers and the individual default thresholds. In the next section, we explain how the optimal posterior distribution from Eq. (3.6) can be used to derive a series of unconditional and conditional probability measures.

3.2.1 Solution of Minimum Cross Entropy

The minimum cross entropy procedure can be viewed as a part of an iterative algorithm to approximate a target probability density f, using empirical data describing its underlying unknown process.[6] In this procedure, an a-priori (or prior) density q is updated to a posterior density p, given the following **Cross Entropy Postulate**:

1. Conditional on a prior density q of a set $\mathfrak{X} \subset \Re^d$,
2. we minimize the Csiszár Cross Entropy measure[7]

$$\mathcal{D}(p \rightarrow q) = \int_{\mathfrak{X}} q(\mathbf{x}) \cdot \psi\left(\frac{p(\mathbf{x})}{q(\mathbf{x})}\right) d\mathbf{x} \tag{3.7}$$

   with respect to $p(\mathbf{x})$, with $\mathbf{x}$ being a column vector and $\mathbf{x} \in \Re^d$,
3. given the moment constraints

$$\mathbb{E}_p K_i(\mathbf{X}) = \int_{\mathfrak{X}} p(\mathbf{x}) \cdot K_i(\mathbf{x}) d\mathbf{x} = \hat{\kappa}_i, i = 0, \ldots, n. \tag{3.8}$$

   where $\{K_i(\mathbf{x})\}_{i=1}^n$ is a set of suitably chosen functions and $\hat{\kappa}_i$ is empirical information describing the behaviour of the system, $\mathbb{E}_f K_i(\mathbf{X})$.

The *Minimum Cross Entropy Problem* is then defined as

$$\min_p \mathcal{D}(p \rightarrow q) \tag{3.9}$$

[6] For further details on the cross-entropy method and its generalizations, please consult with, e.g., Botev and Kroese (2011).

[7] The Csiszár Cross Entropy measure is a measure of directed divergence between probability densities.

subject to the constraints

$$\int_{\mathfrak{X}} p(\mathbf{x}) \cdot K_i(\mathbf{x}) d\mathbf{x} = \hat{\kappa}_i, i = 0, \ldots, n. \tag{3.10}$$

and

$$\int p(\mathbf{x}) d\mathbf{x} = 1 \tag{3.11}$$

The corresponding Lagrangian is then

$$\begin{aligned}
\mathcal{L}(p; \boldsymbol{\lambda}, \lambda_0) = \\
&= \int q(\mathbf{x}) \cdot \psi\left(\frac{p(\mathbf{x})}{q(\mathbf{x})}\right) d\mathbf{x} + \lambda_0 \left(1 - \int p(\mathbf{x}) d\mathbf{x}\right) + \sum_{i=1}^{n} \lambda_i \left(\hat{\kappa}_i - \int p(\mathbf{x}) \cdot K_i(\mathbf{x}) d\mathbf{x}\right) \\
&= \sum_{i=0}^{n} \lambda_i \cdot \hat{\kappa}_i + \int \left(q(\mathbf{x}) \cdot \psi\left(\frac{p(\mathbf{x})}{q(\mathbf{x})}\right) + p(\mathbf{x}) \cdot \sum_{i=0}^{n} \lambda_i \cdot K_i(\mathbf{x})\right) d\mathbf{x},
\end{aligned} \tag{3.12}$$

where $\boldsymbol{\lambda} = [\lambda_1, \lambda_2, \ldots, \lambda_n]^T$, $\hat{\kappa}_0 = 1$, and $K_0(\cdot) = 1$.

Let us assume that $\{K_i(\mathbf{x})\}_{i=0}^n = \{I_i(\mathbf{x})\}_{i=0}^n$, where I_i, $i = 1, 2, \ldots, n$ are binary functions taking values of unity when the respective x_i satisfies some condition, and zero otherwise, and $I_0 = 1$. The first order condition with respect to $p(\mathbf{x})$ is then

$$\frac{\partial \left(q(\mathbf{x}) \cdot \psi\left(\frac{p(\mathbf{x})}{q(\mathbf{x})}\right) + p(\mathbf{x}) \cdot \sum_{i=0}^{n} \lambda_i \cdot \mathbf{I}_i\right)}{\partial p(\mathbf{x})} \stackrel{!}{=} 0 \tag{3.13}$$

the latter can be further simplified as follows:

$$q(\mathbf{x}) \cdot (q(\mathbf{x}))^{-1} \cdot \psi'\left(\frac{p(\mathbf{x})}{q(\mathbf{x})}\right) + \sum_{i=0}^{n} \lambda_i \cdot \mathbf{I}_i = 0 \tag{3.14}$$

$$\psi'\left(\frac{p(\mathbf{x})}{q(\mathbf{x})}\right) = -\sum_{i=0}^{n} \lambda_i \cdot \mathbf{I}_i \tag{3.15}$$

Assume $\psi(\mathbf{x}) = \mathbf{x} \cdot \ln(\mathbf{x})$, which is referred in literature as Kullback–Leibler distance.[8] The Csiszár Cross Entropy measure can then me transformed as

[8] The Kullback–Leibler distance is a usual assumption that allows us to avoid setting additional constraints to secure the non-negativity of $p(\mathbf{x})$.

$$\begin{aligned}\int q(\mathbf{x}) \cdot \psi \left(\frac{p(\mathbf{x})}{q(\mathbf{x})} \right) &= \int q(\mathbf{x}) \cdot \frac{p(\mathbf{x})}{q(\mathbf{x})} \cdot \ln \left(\frac{p(\mathbf{x})}{q(\mathbf{x})} \right) \\ &= \int p(\mathbf{x}) \cdot \ln \left(\frac{p(\mathbf{x})}{q(\mathbf{x})} \right),\end{aligned} \tag{3.16}$$

while our $\psi' \left(\frac{p(\mathbf{x})}{q(\mathbf{x})} \right)$ takes the form

$$\begin{aligned}\psi' \left(\frac{p(\mathbf{x})}{q(\mathbf{x})} \right) &= \left(\frac{p(\mathbf{x})}{q(\mathbf{x})} \cdot \ln \left(\frac{p(\mathbf{x})}{q(\mathbf{x})} \right) \right)' \\ &= \ln \left(\frac{p(\mathbf{x})}{q(\mathbf{x})} \right) + \frac{p(\mathbf{x})}{q(\mathbf{x})} \cdot \left(\frac{p(\mathbf{x})}{q(\mathbf{x})} \right)^{-1} \\ &= \ln \left(\frac{p(\mathbf{x})}{q(\mathbf{x})} \right) + 1,\end{aligned} \tag{3.17}$$

Substituting in our first order condition Eq. (3.15) and simplifying further yields

$$\ln \left(\frac{p(\mathbf{x})}{q(\mathbf{x})} \right) + 1 = -\sum_{i=0}^{n} \lambda_i \cdot \mathbf{I}_i \tag{3.18}$$

$$\ln \left(\frac{p(\mathbf{x})}{q(\mathbf{x})} \right) = -1 - \sum_{i=0}^{n} \lambda_i \cdot \mathbf{I}_i. \tag{3.19}$$

The solution to the Minimum Cross Entropy problem is then

$$p(\mathbf{x}) = q(\mathbf{x}) \cdot exp \left\{ - \left[1 + \sum_{i=0}^{n} \lambda_i \mathbf{I}_i \right] \right\} \tag{3.20}$$

Changing the notation of the Lagrange multiplier of the additivity constraint to μ, we arrive at

$$p(\mathbf{x}) = q(\mathbf{x}) \cdot exp \left\{ - \left[1 + \mu + \sum_{i=1}^{n} \lambda_i \mathbf{I}_i \right] \right\}, \tag{3.21}$$

which is the general form of the solution to the cross-entropy minimization problem.

3.2.2 *Proof of Independence Within the Joint Default Region of the CIMDO Distribution*

This section shows analytically that independence is preserved within the joint probability of distress region of the CIMDO posterior distribution, if multivariate standard normal distribution (the variance-covariance matrix is the identity matrix) is assumed as a prior. The implications of this fact are that the conditional probability of default measure, which uses as an ingredient the joint distress region, equals its unconditional counterpart.

To prove independence, we want to show that using standard normal distribution as a prior, the following holds for the posterior CIMDO distribution and its marginals:

$$P(x_1 > \overline{\mathbf{x}}_1, x_2 > \overline{\mathbf{x}}_2, \ldots, x_n > \overline{\mathbf{x}}_n) = P(x_1 > \overline{\mathbf{x}}_1) \cdot P(x_2 > \overline{\mathbf{x}}_2) \ldots P(x_n > \overline{\mathbf{x}}_n) = \prod_{i=1}^{n} P(x_i > \overline{\mathbf{x}}_i), \tag{3.22}$$

where $P(x_1 > \overline{\mathbf{x}}_1)$, $P(x_2 > \overline{\mathbf{x}}_2)$,..., $P(x_n > \overline{\mathbf{x}}_n)$ and $P(x_1 > \overline{\mathbf{x}}_1, x_2 > \overline{\mathbf{x}}_2, \ldots, x_n > \overline{\mathbf{x}}_n)$ are the cumulative marginal and joint CIMDO probabilities.

Proof We present a *direct* proof of the statement above. We start by expressing $P(x_n > \overline{\mathbf{x}}_n)$, $P(x_1 > \overline{\mathbf{x}}_1, x_2 > \overline{\mathbf{x}}_2, \ldots, x_{n-1} > \overline{\mathbf{x}}_{n-1})$ and $P(x_1 > \overline{\mathbf{x}}_1, x_2 > \overline{\mathbf{x}}_2, \ldots, x_n > \overline{\mathbf{x}}_n)$ in terms of the prior (multivariate standard normal) distribution and the thresholds $\overline{\mathbf{x}}_1$, $\overline{\mathbf{x}}_2$ to $\overline{\mathbf{x}}_n$:

$$\begin{aligned} P(x_n > \overline{\mathbf{x}}_n) &= PoD^n \\ &= \int_{-\infty}^{+\infty}\int_{-\infty}^{+\infty}\cdots\int_{-\infty}^{+\infty} p(x_1, x_2, \ldots, x_n)\mathbf{I}_{[\overline{\mathbf{x}}_n,\infty)} dx_1 \ldots dx_{n-1}dx_n \\ &= \int_{\overline{\mathbf{x}}_n}^{+\infty}\int_{-\infty}^{+\infty}\cdots\int_{-\infty}^{+\infty} (2\pi)^{\frac{n}{2}} e^{\left(-\frac{\sum_{i=1}^{n} x_i^2}{2}\right)} e^{\left(-\left(1+\mu+\sum_{i=1}^{n-1}\lambda_i \mathbf{I}_{[\overline{\mathbf{x}}_i,\infty)}+\lambda_n\right)\right)} dx_1 \ldots dx_{n-1}dx_n, \end{aligned} \tag{3.23}$$

$$\begin{aligned} &P(x_1 > \overline{\mathbf{x}}_1, x_2 > \overline{\mathbf{x}}_2, \ldots, x_{n-1} > \overline{\mathbf{x}}_{n-1}) = \\ &= \int_{-\infty}^{+\infty}\int_{-\infty}^{+\infty}\cdots\int_{-\infty}^{+\infty} p(x_1, x_2, \ldots, x_n)\mathbf{I}_{[\overline{\mathbf{x}}_1,\infty)} \cdot \mathbf{I}_{[\overline{\mathbf{x}}_2,\infty)} \ldots \mathbf{I}_{[\overline{\mathbf{x}}_{n-1},\infty)} dx_1 \ldots dx_{n-1}dx_n \\ &= \int_{-\infty}^{+\infty}\int_{\overline{\mathbf{x}}_{n-1}}^{+\infty}\cdots\int_{\overline{\mathbf{x}}_1}^{+\infty} (2\pi)^{\frac{n}{2}} e^{\left(-\frac{\sum_{i=1}^{n} x_i^2}{2}\right)} e^{\left(-\left(1+\mu+\sum_{i=1}^{n-1}\lambda_i+\lambda_n \mathbf{I}_{[\overline{\mathbf{x}}_n,\infty)}\right)\right)} dx_1 \ldots dx_{n-1}dx_n, \end{aligned} \tag{3.24}$$

where $\mathbf{I}_{[\overline{\mathbf{x}}_1,\infty)}$, $\mathbf{I}_{[\overline{\mathbf{x}}_2,\infty)}$ to $\mathbf{I}_{[\overline{\mathbf{x}}_n,\infty)}$ are indicator functions that take the value of one in the cases where the assets of X_1, X_2 to X_n are beyond their individual thresholds, respectively. Then, the joint probability of distress is as follows:

$$P(x_1 > \overline{\mathbf{x}}_1, x_2 > \overline{\mathbf{x}}_2, \ldots, x_n > \overline{\mathbf{x}}_n) =$$
$$= \int_{-\infty}^{+\infty}\int_{-\infty}^{+\infty} \cdots \int_{-\infty}^{+\infty} p(x_1, x_2, \ldots, x_n)\mathbf{I}_{[\overline{\mathbf{x}}_1,\infty)} \cdot \mathbf{I}_{[\overline{\mathbf{x}}_2,\infty)} \ldots \mathbf{I}_{[\overline{\mathbf{x}}_n,\infty)} dx_1 \ldots dx_{n-1} dx_n$$
$$= \int_{\overline{\mathbf{x}}_n}^{+\infty}\int_{\overline{\mathbf{x}}_{n-1}}^{+\infty} \cdots \int_{\overline{\mathbf{x}}_1}^{+\infty} (2\pi)^{\frac{n}{2}} e^{\left(-\frac{\sum_{i=1}^{n} x_i^2}{2}\right)} e^{\left(-\left(1+\mu+\sum_{i=1}^{n}\lambda_i\right)\right)} dx_1 \ldots dx_{n-1} dx_n, \tag{3.25}$$

Rearranging $P(x_n > \overline{\mathbf{x}}_n)$, we get

$$P(x_n > \overline{\mathbf{x}}_n) =$$
$$= \int_{\overline{\mathbf{x}}_n}^{+\infty} (2\pi)^{-\frac{1}{2}} e^{-\frac{x_n^2}{2}} e^{-\lambda_n} dx_n \int_{-\infty}^{+\infty} \cdots \int_{-\infty}^{+\infty} (2\pi)^{\frac{n-1}{2}} e^{\left(-\frac{\sum_{i=1}^{n-1} x_i^2}{2}\right)} e^{\left(-\left(1+\mu+\sum_{i=1}^{n-1}\lambda_i \mathbf{I}_{[\overline{\mathbf{x}}_i,\infty)}\right)\right)} dx_1 \ldots dx_{n-1}, \tag{3.26}$$

Analogously, for $P(x_1 > \overline{\mathbf{x}}_1, x_2 > \overline{\mathbf{x}}_2, \ldots, x_{n-1} > \overline{\mathbf{x}}_{n-1})$, we come at:

$$P(x_1 > \overline{\mathbf{x}}_1, x_2 > \overline{\mathbf{x}}_2, \ldots, x_{n-1} > \overline{\mathbf{x}}_{n-1}) =$$
$$= \int_{\overline{\mathbf{x}}_{n-1}}^{+\infty} \cdots \int_{\overline{\mathbf{x}}_1}^{+\infty} (2\pi)^{\frac{n-1}{2}} e^{\left(-\frac{\sum_{i=1}^{n-1} x_i^2}{2}\right)} e^{\left(-\sum_{i=1}^{n-1}\lambda_i\right)} dx_1 \ldots dx_{n-1} \int_{-\infty}^{+\infty} (2\pi)^{-\frac{1}{2}} e^{-\frac{x_n^2}{2}} e^{(-(1+\mu+\lambda_n \mathbf{I}_{[\overline{\mathbf{x}}_n,\infty)}))} dx_n, \tag{3.27}$$

Hence, for the product of the latter probabilities, we have:

$$P(x_1 > \overline{\mathbf{x}}_1, x_2 > \overline{\mathbf{x}}_2, \ldots, x_{n-1} > \overline{\mathbf{x}}_{n-1}) \cdot P(x_n > \overline{\mathbf{x}}_n) =$$

$$= \left[\int_{\overline{\mathbf{x}}_{n-1}}^{+\infty} \cdots \int_{\overline{\mathbf{x}}_1}^{+\infty} (2\pi)^{\frac{n-1}{2}} e^{\left(-\frac{\sum_{i=1}^{n-1} x_i^2}{2}\right)} e^{\left(-\sum_{i=1}^{n-1} \lambda_i\right)} dx_1 \ldots dx_{n-1} \int_{-\infty}^{+\infty} (2\pi)^{-\frac{1}{2}} e^{-\frac{x_n^2}{2}} e^{(-(1+\mu+\lambda_n \mathbf{I}_{[\overline{\mathbf{x}}_n,\infty)}))} dx_n \right]$$

$$\cdot \left[\int_{\overline{\mathbf{x}}_n}^{+\infty} (2\pi)^{-\frac{1}{2}} e^{-\frac{x_n^2}{2}} e^{-\lambda_n} dx_n \int_{-\infty}^{+\infty} \cdots \int_{-\infty}^{+\infty} (2\pi)^{\frac{n-1}{2}} e^{\left(-\frac{\sum_{i=1}^{n-1} x_i^2}{2}\right)} e^{\left(-\left(1+\mu+\sum_{i=1}^{n-1} \lambda_i \mathbf{I}_{[\overline{\mathbf{x}}_i,\infty)}\right)\right)} dx_1 \ldots dx_{n-1} \right]$$

$$= \left[\int_{\overline{\mathbf{x}}_n}^{+\infty} \int_{\overline{\mathbf{x}}_{n-1}}^{+\infty} \cdots \int_{\overline{\mathbf{x}}_1}^{+\infty} (2\pi)^{\frac{n}{2}} e^{\left(-\frac{\sum_{i=1}^{n} x_i^2}{2}\right)} e^{\left(-\left(1+\mu+\sum_{i=1}^{n} \lambda_i\right)\right)} dx_1 \ldots dx_{n-1} dx_n \right]$$

$$\cdot \left[\int_{-\infty}^{+\infty} \int_{-\infty}^{+\infty} \cdots \int_{-\infty}^{+\infty} (2\pi)^{\frac{n}{2}} e^{\left(-\frac{\sum_{i=1}^{n} x_i^2}{2}\right)} e^{\left(-\left(1+\mu+\sum_{i=1}^{n} \lambda_i \mathbf{I}_{[\overline{\mathbf{x}}_i,\infty)}\right)\right)} dx_1 \ldots dx_{n-1} dx_n \right]. \tag{3.28}$$

As the integral in the last square brackets is in fact the additivity constraint in our optimization problem, it equals 1 by definition. The remaining term equals our definition for the joint probability $P(x_1 > \overline{\mathbf{x}}_1, x_2 > \overline{\mathbf{x}}_2, \ldots, x_n > \overline{\mathbf{x}}_n)$. If we repeat the procedure iteratively for the joint distributions $P(x_1 > \overline{\mathbf{x}}_1, x_2 > \overline{\mathbf{x}}_2, \ldots, x_i > \overline{\mathbf{x}}_i)$, for $i = n - 1, \ldots, 2$, we arrive at the following decomposition:

$$\begin{aligned} P(x_1 > \overline{\mathbf{x}}_1, x_2 > \overline{\mathbf{x}}_2, \ldots, x_n > \overline{\mathbf{x}}_n) &= P(x_1 > \overline{\mathbf{x}}_1) \cdot P(x_2 > \overline{\mathbf{x}}_2) \ldots P(x_n > \overline{\mathbf{x}}_n) \\ &= \prod_{i=1}^{n} P(x_i > \overline{\mathbf{x}}_i), \end{aligned} \tag{3.29}$$

Hence, the product of the marginal probabilities of distress $P(x_1 > \overline{\mathbf{x}}_1)$, $P(x_2 > \overline{\mathbf{x}}_2)$, …, and $P(x_n > \overline{\mathbf{x}}_n)$ equals the joint probability of distress, meaning that within the joint distress region, the entities X_1, X_2 and X_n are independent. ■

References

Botev, Z., Kroese, D.: The generalized cross entropy method, with applications to probability density estimation. Methodol. Comput. Appl. Probab. **13**, 1–27 (2011)

Gorea, D., Radev, D.: The euro area sovereign debt crisis: can contagion spread from the periphery to the core? Int. Rev. Econ. Financ. **30**, 78–100 (2014)

Hull, J., White, A.: Valuing credit default swaps I: no counterparty default risk. J. Deriv. **8**(1), 29–40 (2000)

Kullback, S.: Information Theory and Statistics. Wiley, Chichester (1959)

Merton, R.: On the pricing of corporate debt: the risk structure of interest rates. J. Financ. **29**(2), 449–470 (1974)

Radev, D.: Assessing systemic fragility: a probabilistic perspective. SAFE Working Paper 70. Research Center SAFE, Goethe University Frankfurt (2014)

Segoviano, M.: Consistent information multivariate density optimizing methodology. FMG Discussion Papers 557. International Monetary Fund (2006)

Chapter 4
Systemic Fragility Measures

4.1 Motivation

The risk of default of sovereigns and banking institutions has proved to be of major concern for regulators during the Global financial and the recent euro area sovereign debt crises. In such turbulent times for markets and regulators, the consistent and assessment of the individual and systemic risks of banks and sovereigns is of utmost importance. As the European Central Bank sets the focus of systemic risk management on the *macrofinancial* level (ECB 2010a), one could conveniently observe the financial system as a portfolio of its institutions, as noted by Lehar (2005). According to Gramlich and Oet (2011), within such a framework the role of a supervisor is twofold. First, to monitor the stability of the portfolio of banks conditional on their individual financial soundness, as well as the soundness of their interlinkages. Second, to analyze and assess the dynamics of the interconnectedness, especially in volatile periods, characterized by extreme events. The authors outline several conditions for a successful systemic risk model: *consistency*, *flexibility*, *forward-looking focus*, *correspondence with empirical data*, *suitability for the need of financial regulators*. In this chapter, we apply the minimum cross-entropy method to develop a number of intuitive measures of systemic risk of sovereigns and banks to complement the existing toolkit of regulators.

The reason why we investigate euro area sovereign default parallel to banking default is straightforward. At the wake of the sovereign debt crisis EA banks had large exposures to EA government debt (see, e.g., ECB 2010b; EBA 2011; IMF 2011; Acharya and Thakor 2016), hence a possible negative shock from sovereigns to the banking sector may cause a collapse of the whole EA financial system. Sovereign default, however rare, has serious welfare costs not only for the parties involved in the debt contract, but also for third parties. On the one side, we have a loss of reputation and limited future access to international debt markets of the defaulting sovereign (see, e.g., Panizza et al. 2009). On the other, the default has direct negative effect on domestic or foreign private and public investors, who, in the worse case

D. Radev, *Measuring Systemic Risk*, Studies in Systems, Decision and Control 409,
https://doi.org/10.1007/978-3-030-94281-6_4

scenario, could become distressed as well (see, e.g., Arteta and Hale 2008). Apart from financial channels, contagion can also spread through real economy channels, as well, because of the strong economic links not only in the common currency area, but also within the European Union (EU) and globally (Gorea and Radev 2014).

The chapter is organized as follows. In Sect. 4.2, we introduce our probability measures and provide guidelines for their calculation. Section 4.3 describes briefly our dataset, while Sect. 4.4 presents our empirical results. Section 4.5 presents detailed robustness checks with respect to some of the main parameters underlying the cross-entropy approach: the prior distribution, the dependence structure assumptions, and the performance in the default region of the joint distribution. Section 4.6 concludes.

4.2 Probability Measures

4.2.1 Individual Probabilities

We apply the bootstrapping procedure and use all available maturities from 1 to 5 year of CDS spreads to recover the PoD of a bank or a sovereign. The CDS contracts have quarterly premium payments as a general rule, so we adjust the procedure accordingly. We also correct for accrual interest, as suggested by Adelson et al. (2004). As risk-free rates, required as inputs, we use all available maturities of AAA Euro Area bond yields from 1 to 5 years. The recovery rate is uniformly set at 40%, both for banks and sovereigns, as this is the prevailing assumption in literature and practice.[1] The resulting series are cumulative probabilities of default and to accommodate the one-year horizon of interest to policy makers, they have to be annualized, using the formula:

$$PoD_t^{annual} = 1 - (1 - PoD_t^{cum})^{\frac{1}{T}}, \tag{4.1}$$

where T is the respective time horizon (T = 5 for 5-year PoD) and PoD_t^{annual} is the annualized version of the cumulative PoD_t^{cum}.

Figure 4.1 presents the results from both procedures for a distressed sovereign, namely Greece, for the period 01.01.2007–31.12.2011. We notice the main drawback of the simple calculation method—while the series generally overlap in tranquil times, they diverge during the distress period starting from May 2010. The margin

[1] Sturzenegger and Zettelmeyer (2008) find that the historical sovereign recovery rates are usually between 30 and 70%. Lucas et al. (2014) use those results as motivation to choose 25% recovery rate for their default estimations. We decide to be more conservative with regard to the loss given default assumption, as the recent negotiations for the Private Sector Involvement (PSI) in the Greek bailout packages suggest haircuts between 50 and 70%. As non-institutional investors are the main participants in the CDS markets, we argue that their expectations of default risk are what the CDS spreads reflect, thus we remain with the usual recovery rate convention in financial literature. For a discussion on how different recovery rates affect the PoD estimates, please refer to our robustness checks Sect. 4.5 and to Gorea and Radev (2014).

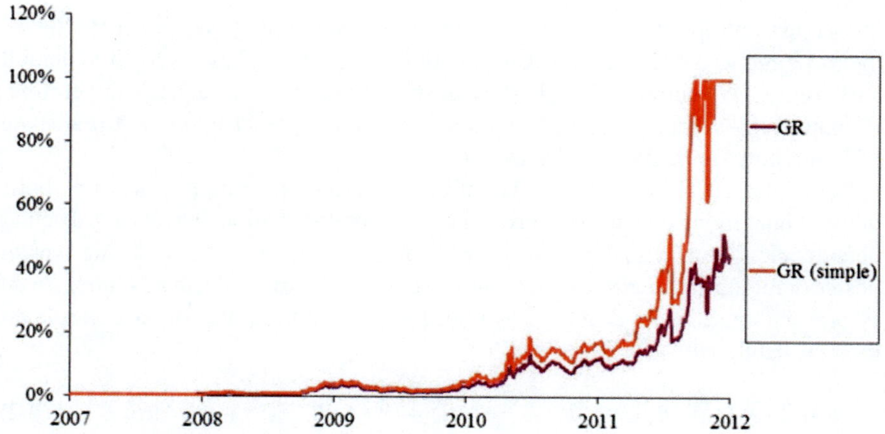

Fig. 4.1 5-year annualized CDS-implied probabilities of default of Greece, using the simple formula Eq. (3.1) (GR(simple)) and the bootstrapping procedure (GR). The 5-year annualized CDS-implied bootstrapped probabilities of default are derived from the respective cumulative ones using formula Eq. (4.1). Euro-denominated CDS spreads are used. Period: 01.01.2007–31.12.2011. *Source* own calculations

increases rapidly with the rise of CDS spreads, leading to results higher than unity at the end of the period, which we truncate at 1 to match the definition of probability. The bootstrapped probabilities, on the other hand, have fairly reasonable annualized values in the distress period, peaking in the region from 45 to 50%. The reason for this misalignment is that the naïve Equation (3.1) can be seen as a linear approximation of the more elaborate bootstrapping procedure, and does not account for all its caveats. The formula performs well for entities with low levels of CDS spreads (Germany, France, Banco Santander), but fails for distressed sovereigns or corporates (Greece, Dexia).

As shown in the previous chapter, once we have specified the default regions of the *prior* distribution and recovered the optimal Lagrange multipliers, we can completely identify the *posterior* distribution. Moreover, by coupling the different prior default regions with their respective Lagrange multiplier combinations, we can fairly easy construct a practically unlimited number of default probability measures. The current section aims at presenting this variability of our methodology.

4.2.2 Systemic Fragility Measure

Our first measure is the probability of *at least* two entities (sovereigns or banks) defaulting jointly. This is an unconditional measure and represents the systemic default *potential*, and hence the general vulnerability to systemic events. In other words, it is an indicator of how *fragile* the system is to default of its constituents. Therefore, we call this indicator the Systemic Fragility Measure (SFM). In essence, it is a cross-entropy-derived alternative to a similar measure by Avesani et al. (2006), the Systemic Risk Indicator for banks, with the latter being based on *n*th to default baskets of CDS contracts. Such baskets are not traded for sovereigns, and therefore, in contrast to Avesani et al. (2006), our cross-entropy approach allows us to calculate SFM not only for banks, but also for sovereigns.

As explained above, the calculation of the various probability measures boils down to integrating over the regions of the posterior distribution, where the particular default cases occur. With regard to the SFM, this means that we have to sum up over all posterior regions, where at least two entities default. For example, in a 3-dimensional system with entities A, B and C,[2] we need to sum up the following *unconditional* joint probabilities:

$$SFM = P(\neg A, B, C) + P(A, \neg B, C) + P(A, B, \neg C) + P(A, B, C), \quad (4.2)$$

where the "¬" sign indicates that the respective entity does not default.

The remaining subsections introduce *conditional* probability measures that aim to capture *complex* scenarios—the effect of the default of a particular entity or subset of entities on the distress risk vulnerability of the remaining institutions in the financial system.

4.2.3 Probability of A Defaulting Given B Defaults

We start with the simplest extension beyond the *unconditional* joint probability framework: the probability of default of entity A given entity B defaults ($P(A|B)$). Deriving $P(A|B)$ is a direct application of the Bayes' rule:

$$P(A|B) = \frac{P(A, B)}{P(B)}, \quad (4.3)$$

where $P(A, B)$ is the *joint* probability of default of entities A and B, while *P(B)* is the *marginal* probability of default of entity B.

Note that in a system with N entities, the calculation of this bivariate measure effectively involves decreasing the dimensionality from N to 2 (for the calculation of $P(A, B)$) and to 1 (for the calculation of $P(B)$). Both ingredients of the conditional probability (4.3) could be derived by integrating the multivariate joint density over the remaining N − 2 or N − 1 entities. Intuitively, an integration over N − 1 entities is in fact equivalent to using the CDS-derived marginal probabilities of default of the Nth entity. Therefore, only the integration over the N − 2 entities not involved in the respective conditional probability suffices for the calculation of Eq. (4.3).

---

[2] The extension to the multivariate case is trivial.

4.2.4 Probability of A Defaulting Given B and C Default

The next indicator measures the conditional probability of default of an entity, given two other entities default simultaneously. In the Bayes' framework, mentioned above, this probability of default is defined as

$$P(A|B,C) = \frac{P(A,B,C)}{P(B,C)}, \tag{4.4}$$

with $P(A, B, C)$ and $P(B, C)$ being, respectively, the joint probabilities of entities A, B and C, and of entities B and C defaulting.

The procedure for the calculation of the measure is similar to the method in the previous subsection, but this time it involves decreasing the N-dimensional joint probability of default to its 3- and 2-dimensional versions.

4.2.5 Probability of at Least N − 1 Additional Entities Defaulting Given Entity A Defaults

Our final (and most complex) probability measure is the probability of at least N − 1 entities (banks or sovereigns) defaulting, given a particular entity defaults.[3] This measure is a generalization of the probability of at least one (PAO) bank defaulting,[4] introduced in Segoviano and Goodhart (2009) and aims at gauging the expected severity of a crisis stemming from a particular entity, and hence, the rate of *contagion penetration* in the financial system.

To define the measure, let us consider again a system of three banks, A, B and C.[5] The probability of at least one additional bank defaulting given a particular bank (say C) defaults is then

$$PNBD(\text{at least } 1|C) = P(A|C) + P(B|C) - P(A,B|C), \tag{4.5}$$

where $P(A|C)$, $P(B|C)$ and $P(A, B|C)$ are the respective conditional probabilities for all possible default contingencies. Using this intuition, it is easy to proceed one step further and derive the probability of at least *two* banks (in this case A and B) defaulting given bank C defaults:

$$PNBD(\text{at least } 2|C) = P(A,B|C). \tag{4.6}$$

[3] In the text, we abbreviate the measure related to banks as PNBD and the one related to sovereigns as PNSD.

[4] PAO is also sometimes referred to as a probability of spillover effects (PSE).

[5] The extension to higher dimensions, although more involving, is straightforward, as long as one keeps account of the default contingencies to be added or subtracted.

Table 4.1 List of euro area large and complex banking institutions used in our analysis

Euro area large and complex banking groups

	Country code	Name
1	BE	Dexia SA
2	DE	Bayerische Landesbank
3	DE	Commerzbank AG
4	DE	Deutsche Bank AG
5	DE	DZ Bank AG
6	DE	Landesbank Baden-Württemberg
7	ES	Banco Bilbao Vizcaya Argentaria
8	ES	Banco Santander SA
9	FR	BNP Paribas
10	FR	Credit Agricole SA
11	FR	Societe Generale
12	IT	Intesa Sanpaolo SpA
13	IT	UniCredit SpA
14	NL	ING Groep NV
15	NL	Rabobank

In the limit (i.e. for N-1 additional entities defaulting) the PNBD/PNSD converges to the Conditional Joint Probability of Default (CoJPoD), introduced in Radev (2019):

$$CoJPoD(System_{-C}|C) = P(System_{-C}|C), \tag{4.7}$$

where $CoJPoD(System_{-C}|C)$ is the probability of the remaining members of the system (banks, sovereigns or both), to default, given a bank or sovereign C defaults.

4.3 Data

We recover marginal probabilities of default using CDS premia for contracts with maturities from 1 to 5 years for the period 01.01.2007 and 31.12.2011. The bootstrapping procedure requires as additional inputs refinancing interest rates, which we choose to be the AAA euro area government bond yields for maturities from 1 to 5 years. The CDS spreads and the government bond yields are at daily frequency, which is also the frequency of the resulting probabilities of default. Our analysis covers 10 euro area sovereigns (Austria, Belgium, France, Germany, Greece, Ireland, Italy, Netherlands, Portugal and Spain) and 15 EA large and complex banking groups.[6] Each banking group has a minimum total assets value of EUR 200 Billion.

[6] The banks used in our analysis are listed in Table 4.1.

Table 4.2 Descriptive statistics of the 5-year CDS spread series of 15 LCBGs: BBVA, Santander, BayernLB, BNP PARIBAS, Commerzbank, Credit Agricole, Deutsche Bank, Dexia, DZ Bank, ING Groep, Intesa Sanpaolo, Landesbank BW, Rabobank, Societe Generale and UniCredit. The data are in basis points. Period: 01.01.2007–31.12.2011

	BBVA	B. Santander	BayernLB	BNP Paribas	Commerzbank
Minimum	7.10	7.00	10.52	5.50	7.90
Mean	133.65	130.55	111.88	82.82	101.93
Maximum	473.17	473.54	348.82	361.62	364.65
Std. dev.	98.14	94.41	66.23	63.02	67.10
Nr. of obs.	1305	1305	1305	1305	1305
	Credit Agricole	Deutsche Bank	Dexia	DZ Bank	ING Groep
Minimum	5.50	8.70	6.50	12.92	4.00
Mean	103.56	94.39	244.41	88.46	94.42
Maximum	363.31	317.28	962.31	186.50	251.79
Std. dev.	67.29	49.69	188.58	37.11	54.76
Nr. of obs.	1305	1305	1305	1305	1305
	Intesa Sanpaolo	Landesbank BW	Rabobank	Societe Generale	UniCredit
Minimum	5.40	1.00	2.50	5.70	7.48
Mean	111.53	106.93	70.33	107.96	103.49
Maximum	625.50	330.40	204.30	434.14	398.89
Std. dev.	108.98	78.21	42.00	82.24	70.80
Nr. of obs.	1305	1305	1305	1305	1305

Table 4.2 presents the descriptive statistics of the 5-year CDS spreads of the 15 banks in our sample. The average 5-year CDS spread in the cross-section ranges from 70.3 basis points for Rabobank to 244.4 basis points for Dexia. We also notice a substantial increase of CDS premia even for the safest bank in the beginning of the sample, Landesbank Baden-Württemberg, from 1 to 330.4 basis points. However, this does not compare to the dynamics of the price for protection against the default of Dexia, which starts at 6.5 basis points at the beginning of the period and reaches a maximum almost 1000 basis points. We also notice that, on average, French, German and Dutch banks exhibit the lowest volatility in the price for protection against default.

Table 4.3 reports the descriptive statistics of the CDS spreads of the ten sovereigns. The spreads range from 0.5 basis points for Austria and France almost 14400 basis points for Greece. We also notice a substantial increase of CDS premia even for the safest country, Germany, from 4.5 to 121 basis points. The sovereigns with the lowest volatility are France, Germany and the Netherlands, which mirrors the data for the banks in these countries. Overall, we observe that, while the sovereigns spreads are much lower than bank spreads in the beginning of the sample, their maximum values and volatility is higher for the overall period.

Table 4.3 Descriptive statistics of the 5-year CDS spread series of Austria (AT), Belgium (BE), France (FR), Germany (GE), Greece (GR), Ireland (IE), Italy (IT), the Netherlands (NL), Portugal (PT), Spain (ES). The data are in basis points. Period: 01.01.2007–31.12.2011

	AT	BE	FR	GE	GR
Minimum	0.50	1.40	0.50	0.60	4.40
Mean	66.45	84.73	52.28	32.26	843.07
Maximum	273.00	403.01	247.08	120.59	14395.72
Std. dev.	56.98	84.09	52.14	26.74	1780.04
Nr. of obs.	1305	1305	1305	1305	1305
	IE	IT	NL	PT	ES
Minimum	1.75	5.30	1.00	3.40	2.47
Mean	259.38	128.37	38.29	254.63	133.88
Maximum	1286.91	595.68	136.21	1308.51	490.86
Std. dev.	277.35	124.46	32.50	342.73	123.42
Nr. of obs.	1305	1305	1305	1305	1305

Tables 4.4 and 4.5 present the empirical correlation matrices of the bilateral changes in the respective 5-year CDS spreads for the sample of banks and sovereigns, respectively. We use these matrices to define the dependence structure of the prior distribution for the cross-entropy approach. Overall, we observe high positive values for the bilateral correlations, which is an initial indicator that the assets (and defaults) of banks are perceived to be correlated. This provides a strong support for augmenting the approach of Segoviano (2006) and to use empirical correlations, instead of assuming zero correlation among banks and sovereigns. This point is explored further in the robustness checks in Sect. 4.5.

4.4 Empirical Results

4.4.1 Marginal Probability of Default Results

This section presents the results for the individual probabilities of default for our samples of banks and sovereigns.

Individual Bank Results

In Figs. 4.2 and 4.3, we present the 5-year annualized CDS-implied probabilities of default of the 15 banks in our sample. The series reveal that the LCBGs were considered relatively default-free before the outbreak of the subprime crisis and

Table 4.4 Correlation structure between 15 LCBGs: BBVA, Santander, BayernLB, BNP PARIBAS, Commerzbank, Credit Agricole, Deutsche Bank, Dexia, DZ Bank, ING Groep, Intesa Sanpaolo, Landesbank BW, Rabobank, Societe Generale and UniCredit. Period: 01.01.2007–31.12.2011. The correlations are calculated between changes in the 5-year CDS spreads of the sovereigns in the respective column and row

	BBVA	B. Santander	BayernLB	BNP	Commerz.	Cr. Agr.	Deutsche	Dexia	DZ	ING	Intesa	LBBW	Rabo	Soc. Gen.	UniCredit
BBVA	1.00	0.82	0.41	0.78	0.74	0.82	0.76	0.50	0.27	0.54	0.80	0.27	0.63	0.76	0.67
Santander		1.00	0.39	0.69	0.65	0.71	0.67	0.45	0.25	0.56	0.73	0.23	0.58	0.68	0.61
BayernLB			1.00	0.46	0.47	0.45	0.42	0.33	0.20	0.38	0.43	0.33	0.38	0.47	0.39
BNP				1.00	0.78	0.90	0.80	0.51	0.30	0.57	0.84	0.27	0.63	0.91	0.65
Commerz.					1.00	0.78	0.81	0.50	0.27	0.56	0.75	0.32	0.61	0.76	0.61
Cr. Agr.						1.00	0.78	0.52	0.31	0.58	0.84	0.26	0.62	0.92	0.68
Deutsche							1.00	0.51	0.27	0.62	0.77	0.26	0.69	0.77	0.70
Dexia								1.00	0.17	0.44	0.54	0.18	0.46	0.50	0.43
DZ									1.00	0.19	0.32	0.30	0.19	0.32	0.21
ING										1.00	0.57	0.20	0.61	0.57	0.59
Intesa											1.00	0.29	0.60	0.83	0.70
LBBW												1.00	0.20	0.28	0.24
Rabo													1.00	0.59	0.63
Soc. Gen.														1.00	0.63
UniCredit															1.00

Table 4.5 Correlation structure between 10 sovereigns: Austria (AT), Belgium (BE), France (FR), Germany (GE), Greece (GR), Ireland (IE), Italy (IT), Netherlands (NL), Portugal (PT), Spain (ES). Period: 01.01.2007–31.12.2011. The correlations are calculated between changes in the 5-year CDS spreads of the sovereigns in the respective column and row

	AT	BE	FR	GE	GR	IE	IT	NL	PT	ES
AT	1.00	0.70	0.73	0.75	0.13	0.54	0.66	0.79	0.46	0.62
BE		1.00	0.82	0.74	0.16	0.69	0.83	0.72	0.65	0.81
FR			1.00	0.82	0.22	0.63	0.81	0.73	0.60	0.76
GE				1.00	0.19	0.59	0.72	0.75	0.56	0.69
GR					1.00	0.19	0.21	0.16	0.17	0.17
IE						1.00	0.71	0.55	0.77	0.74
IT							1.00	0.68	0.71	0.90
NL								1.00	0.49	0.63
PT									1.00	0.73
ES										1.00

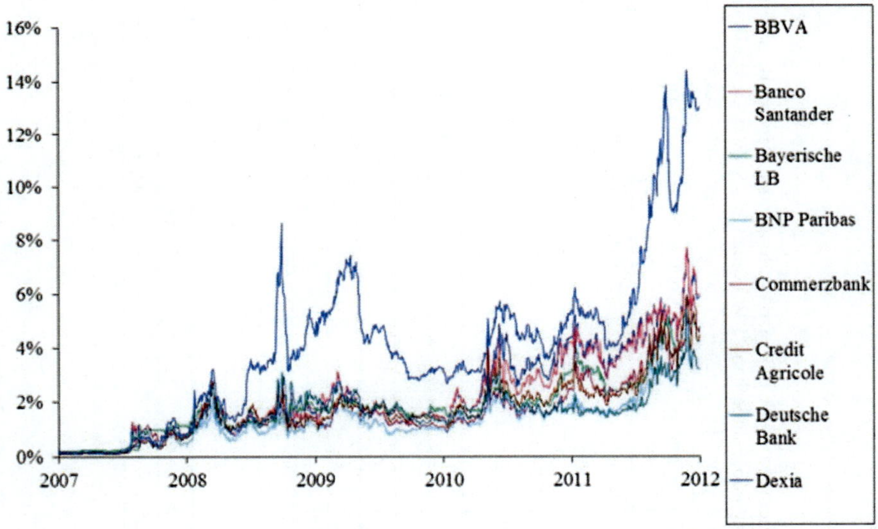

Fig. 4.2 5-year annualized CDS-implied bootstrapped probabilities of default for 8 banks: BBVA, Banco Santander, Bayerische LB, BNP Paribas, Commerzbank, Credit Agricole, Deutsche Bank and Dexia. Euro-denominated CDS spreads are used. Period: 01.01.2007–31.12.2011. *Source* own calculations

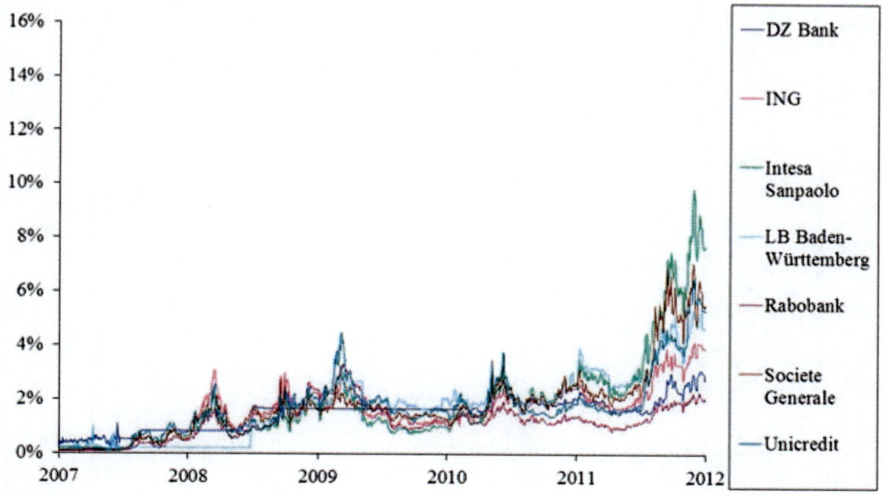

Fig. 4.3 5-year annualized CDS-implied bootstrapped probabilities of default for 7 banks: DZ Bank, ING, Intesa Sanpaolo, LB Baden-Württemberg, Rabobank, Societe Generale and Unicredit. Euro-denominated CDS spreads are used. Period: 01.01.2007–31.12.2011. *Source* own calculations

experienced the first peak of marginal default risk before the bailout of Bear Stearns in the spring of 2008. We notice that, with the exception of Dexia, the individual default risk was stable around 2% during the financial crisis and the following global recession. Two subperiods can be discerned from the PoD dynamics in the second half of the sample period: between the exacerbation of the sovereign debt crisis at the end of the first quarter of 2010 and the second quarter of 2011, and the period thereafter.

Individual Sovereign Results

The sovereign marginal PoD results are presented in Fig. 4.4. We observe that before Lehman Brothers' debacle, EA sovereign bonds were perceived as riskless. The individual default risk rises in the following global recession, but still remains at relatively low levels. After the outbreak of the sovereign debt crisis in late 2009 and early 2010, we observe strong divergence in investors' perceptions about EA sovereigns' individual default risk. Apart from the outlier Greece, we can distinguish several other "groups" of countries that are perceived equally risky by the end of the sample period: Ireland and Portugal; Belgium, Italy and Spain; and Austria, France, Germany and Netherlands.

Our next sections present the results from our multivariate analysis, based on the cross-entropy approach.

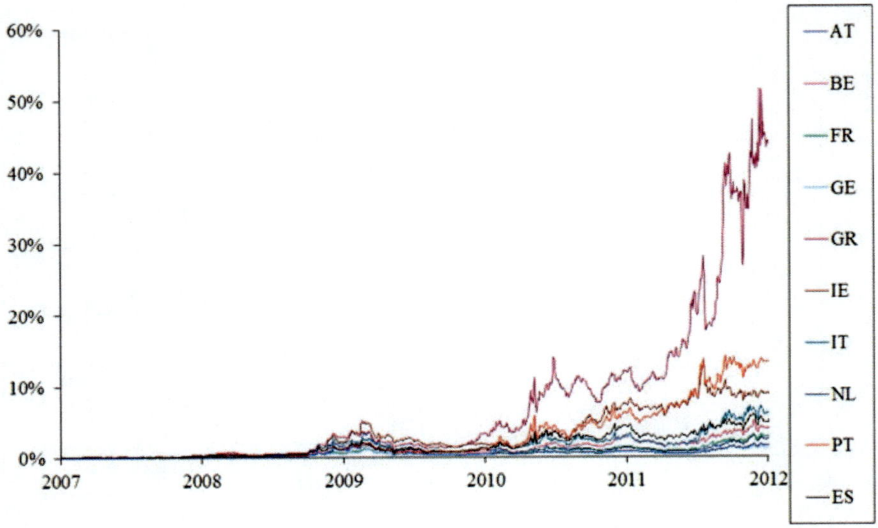

Fig. 4.4 5-year annualized CDS-implied bootstrapped probabilities of default for 10 sovereigns: Austria (AT), Belgium (BE), France (FR), Germany (GE), Greece (GR), Ireland (IE), Italy (IT), Netherlands (NL), Portugal (PT), Spain (ES). Euro-denominated CDS spreads are used. Period: 01.01.2007–31.12.2011. *Source* own calculations

4.4.2 Systemic Fragility Measure Results

Figures 4.5, 4.6, 4.13 and 4.14 present the banking and sovereign Systemic Fragility Measures, plotted against major regulatory and systemic events since the beginning of 2007. The results for the 15 LCBGs show a significant increase in the EA banking system fragility since the onset of the subprime crisis. The various rescue packages throughout the period seem to temporarily relieve the default pressure in our banking sample. The joint default risk in our 15-dimensional system oscillates around 5% for a large part of the sample period, but sharply increases after the bailout request by the Portuguese government in the spring of 2011. The negative investor default risk sentiments continue until the end of the sample and reach its maximum soon after the private sector involvement (PSI) agreement for Greece in October 2011. Overall, we observe that the market joint default expectations seem to react positively to regulatory intervention announcements, but these effects appear to be of *transitory nature*.

In Fig. 4.14, we depict the SFM results for the 10 sovereigns in our sample.[7] Until Lehman Brothers' debacle, markets did not expect that the subprime crisis would affect the solvency of EA governments. The sovereign SFM increases significantly during the global recession period in the beginning of 2009, but subsides gradually thereafter. Although the levels of the banking and sovereign SFM cannot be compared directly, one should notice the much steeper increase of sovereign systemic risk

[7] Note that the levels of SFM between Figs. 4.13 and 4.14 are not directly comparable, because of the different number of dimensions of the underlying distributions.

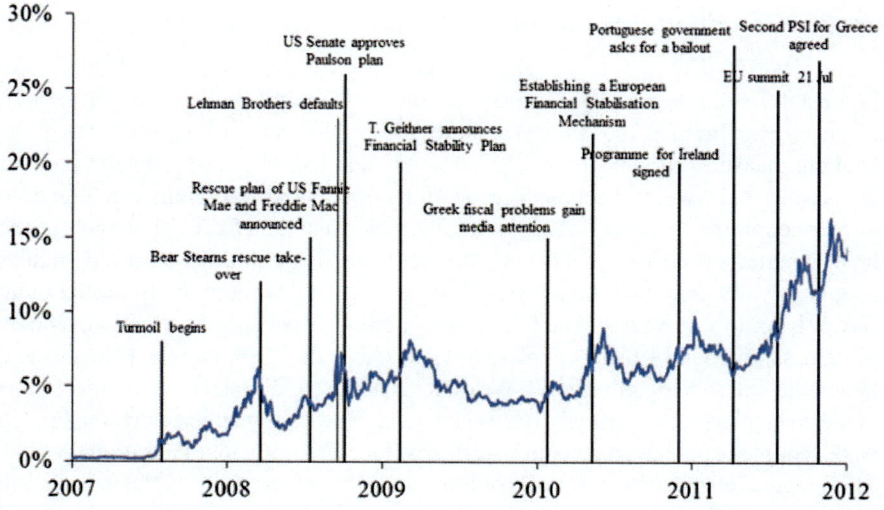

Fig. 4.5 Bank Systemic Fragility Measure (probability of at least two *banks* jointly defaulting) involving 15 large and complex banking groups in the period 01.01.2007–31.12.2011. The measure is derived incorporating empirical correlation, calculated between changes of the respective banks' 5-year CDS spreads. *Source* own calculations

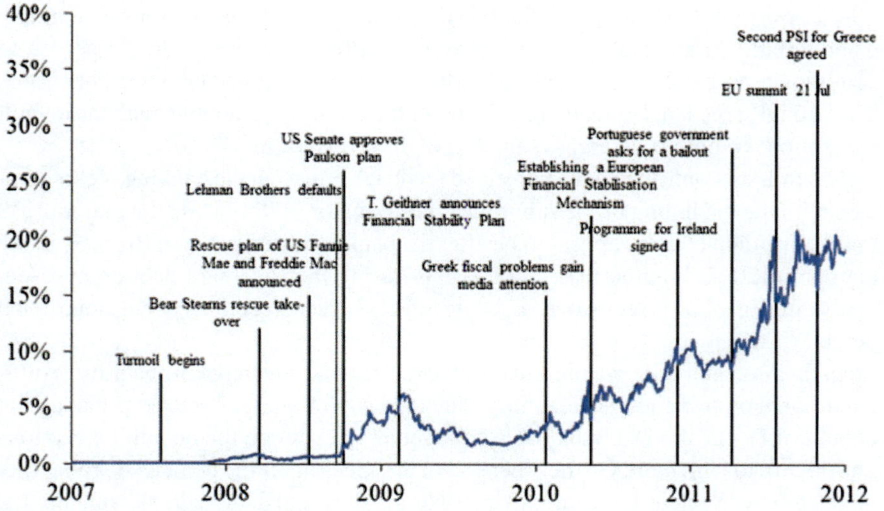

Fig. 4.6 Sovereign Systemic Fragility Measure (probability of at least two *sovereigns* jointly defaulting) involving 10 EA countries in the period 01.01.2007–31.12.2011. The measure is derived incorporating empirical correlation, calculated between changes of the respective sovereigns' 5-year CDS spreads. *Source* own calculations

from the beginning of 2010 on. We document the same temporary positive effect of regulatory interventions as in the case of the banks sample. Even the sharp decline in risk following the EU summit on 21 July 2011 is compensated for by the end of the year.

4.4.3 Conditional Probabilities Results

Figures 4.7 and 4.8 depict the univariate bank (sovereign) probabilities of default, given a particular bank (sovereign) defaults.[8] We notice very different patterns in the banking measures. It appears that a distressed bank like Dexia (throughout most of the period) had lower impact on the probability of default perceptions with regard to a safer counterpart (Intesa Sanpaolo or Unicredit) than the effect of a default of the latter counterpart on Dexia. This is intuitive, since if the safer bank in a particular couple actually defaulted, investors would expect that it is more likely for the more riskier bank to follow suit. The bottom two plots deserve special attention, as they present same-country bank couples: Societe Generale–BNP Paribas (France) and Unicredit–Intesa Sanpaolo (Italy). We notice that the conditional default perceptions of the particular banks narrowly trace each other. However, it appears that the French banks couple is perceived generally as the riskier one throughout the sample period. This is again related to the low PoDs of French banks, but also to the higher correlation between these banks' assets: 0.91, compared to 0.70 between Intesa Sanpaolo and UniCredit.

Turning to the sovereign couples (Fig. 4.8), we observe that investors perceive Greece to be very sensitive to default expectations concerning other EA countries (Portugal and Spain). While we find a reciprocal effect for Portugal, given an expected Greek government default, the effect on the Spanish default perceptions is minimal. Conditioning on a default of Spain provides a different picture, with the perceived conditional probability of default of Portugal rising steadily since November 2009. The strongest feedback effects are present for the Italy–Spain couple, with the default perceptions of both sovereigns being tightly linked since mid-2007.

Overall, we confirm that the perceived feedback effects in the banking sector coincide with the beginning of the subprime crisis (August 2007), while the major peaks in the sovereign couples appear to be after the bankruptcy of Lehman Brothers in the autumn of 2008. In either cases, the events during the sovereign debt crisis, especially after July 2011, seem to have strong effect on the perceptions about conditional default risk.

In Figs. 4.9 and 4.10, we present univariate bank and sovereign probability results, conditional on *two* entities defaulting. The top subplot in Fig. 4.9 reveals that a joint default of Dexia and DZ Bank has a relatively low effect on the default perceptions with regard to Unicredit. On the other end of the spectrum is the perceived probability of default of Societe Generale given BNP Paribas and Unicredit default jointly. Considering Deutsche Bank, international investors seem to expect that it is less vulnerable to Dexia and DZ bank defaulting than to a joint default of Societe Generale and Unicredit.

Turning to perceptions about sovereign default given two states go bankrupt (Fig. 4.10), we note that investors expect Spain to react relatively more intensively to a default of Italy and Portugal (second plot from the top) than to a simultaneous

[8] For the sake of brevity, we report the results only for several of the couples. The remaining results are available upon request.

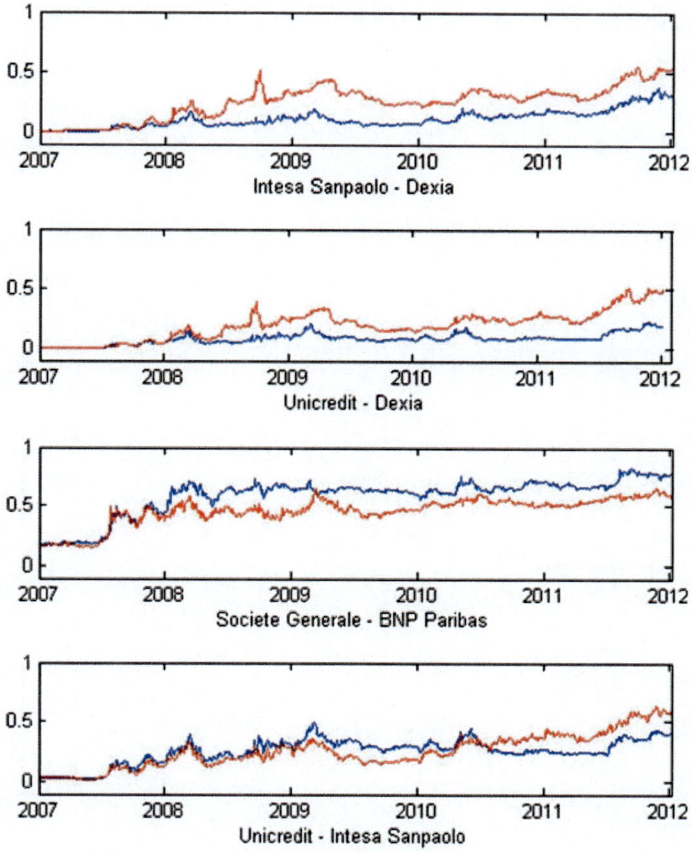

Fig. 4.7 Banking conditional probability of default given a particular bank defaults: 5-year annualized conditional probabilities of default of selected complex banking groups in the period 01.01.2007–31.12.2011. The blue (red) line corresponds to the probability of default of the first (second) bank listed in the couple, given the second (first) bank defaults. E.g. the blue line in the top plot represents the probability of default of Intesa Sanpaolo given Dexia defaults, while the red line corresponds to the probability of default of Dexia given Intesa Sanpaolo defaults. The probabilities derivation incorporates empirical correlation, calculated between changes of the respective banks' 5-year CDS spreads. *Source* own calculations

default of Greece and Portugal. Nonetheless, both contingencies seem to have a very high overall effect on the expectations about a Spanish default. The same holds for the perceptions about Italy defaulting given that Portugal and Spain jointly default (second plot from the bottom). Here, we notice a significant increase in the expected Italian conditional probability of default since mid-2011. Finally, the last plot in Fig. 4.10 shows that the joint default of Italy and Spain has limited effect on the expected probability of default of Germany.

Our last measure, the probability of at least N-1 additional entities defaulting given a particular entity defaults, is presented in Fig. 4.11 (for banks) and Fig. 4.12

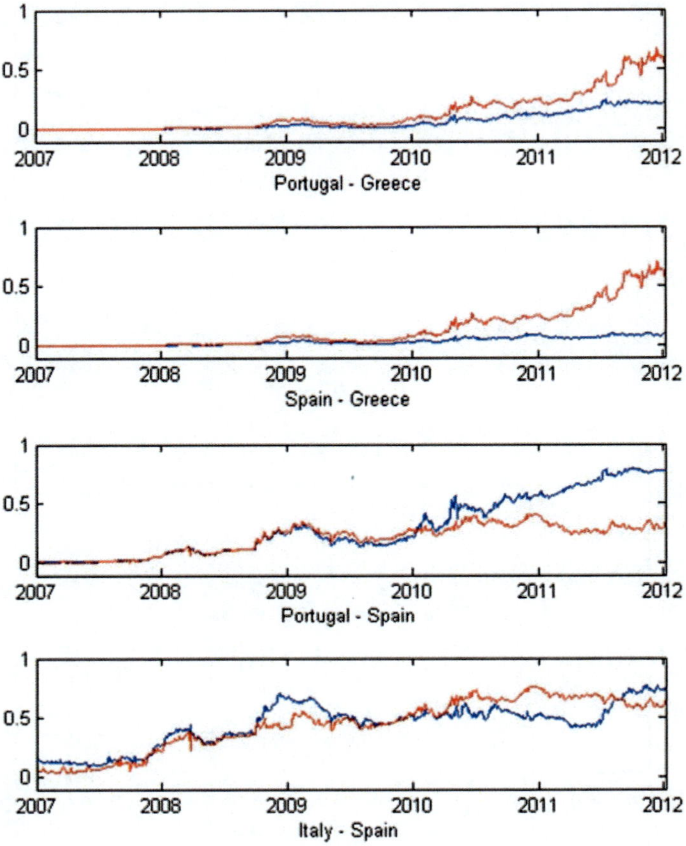

Fig. 4.8 Sovereign conditional probability of default given a particular country defaults: 5-year annualized conditional probabilities of default of selected sovereigns in the period 01.01.2007–31.12.2011. The blue (red) line corresponds to the probability of default of the first (second) country listed in the couple, given the second (first) state defaults. E.g. the blue line in the top plot represents the probability of default of Portugal given Greece defaults, while the red line corresponds to the probability of default of Greece given Portugal defaults. The probabilities derivation incorporates empirical correlation, calculated between changes of the respective sovereigns' 5-year CDS spreads. *Source* own calculations

(for sovereigns). As mentioned before, for convenience, we call the measure for banks PNBD and the one for sovereigns PNSD. An additional detail is that, for presentation purposes, each curve in Figs. 4.11 and 4.12 presents the cross-sectional median values of the respective probability. Although we still cannot compare the magnitudes of the probabilities of banks and sovereigns, we would expect that if the dimensions of the system's distribution are higher, there would be a higher likelihood of at least one entity defaulting given another one defaults. Therefore, in general, the conditional probability values for the 15-dimensional banks distribution would be higher than the corresponding values for the 10-dimensional sovereign distribution.

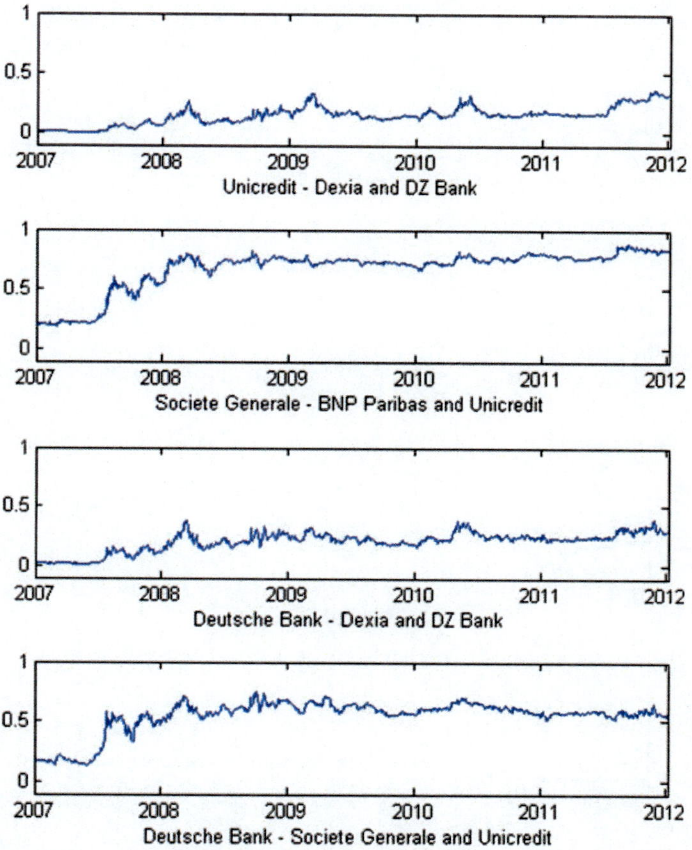

Fig. 4.9 Banking conditional probability of default given two banks default: 5-year annualized conditional probabilities of default of selected complex banking groups in the period 01.01.2007–31.12.2011. The blue line corresponds to the probability of default of the first bank listed in the couple, given the remaining two listed banks default simultaneously. E.g. the blue line in the top plot represents the probability of default of Unicredit given Dexia and DZ Bank default. The probabilities derivation incorporates empirical correlation, calculated between changes of the respective banks' 5-year CDS spreads. *Source* own calculations

The results confirm the impression from all our previous measures that the distress in the banking system started already in mid-2007, while we witness a slow build-up of sovereign systemic risk up to the default of Lehman Brothers. A surprising finding is that, at least when medians are compared, the probabilities of default of at least 1, 2 and 3 sovereigns given a particular sovereign defaults are higher than the corresponding PNBD values for most of the second half of the sample period. What can be also noted is that the conditional probability of *at least one* entity defaulting rises very fast to the unity limit of the probability domain (the unreported maximum values are even closer to 1), making the dynamics of this measure (introduced in Segoviano and Goodhart 2009; and often referred to as the probability of spill-

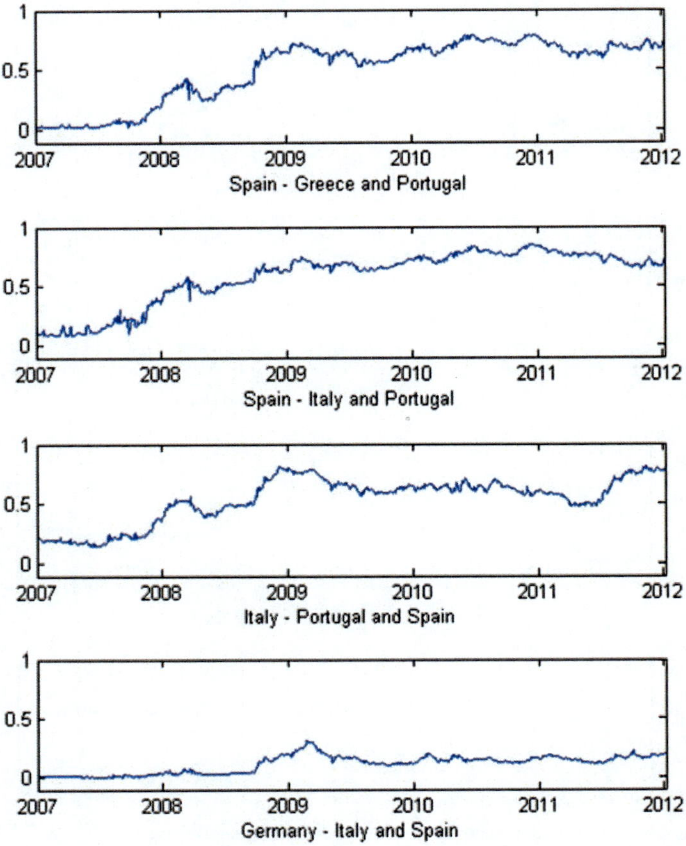

Fig. 4.10 Sovereign conditional probability of default given two states default: 5-year annualized conditional probabilities of default of selected sovereigns in the period 01.01.2007–31.12.2011. The blue line corresponds to the probability of default of the first sovereign listed in the couple, given the remaining two listed sovereigns default simultaneously. E.g. the blue line in the top plot represents the probability of default of Spain given Greece and Portugal default. The probabilities derivation incorporates empirical correlation, calculated between changes of the respective sovereigns' 5-year CDS spreads. *Source* own calculations

over effects) relatively uninformative. Therefore, we believe that our generalization provides a richer picture of the depth of penetration of default spill-over effects within the financial system.

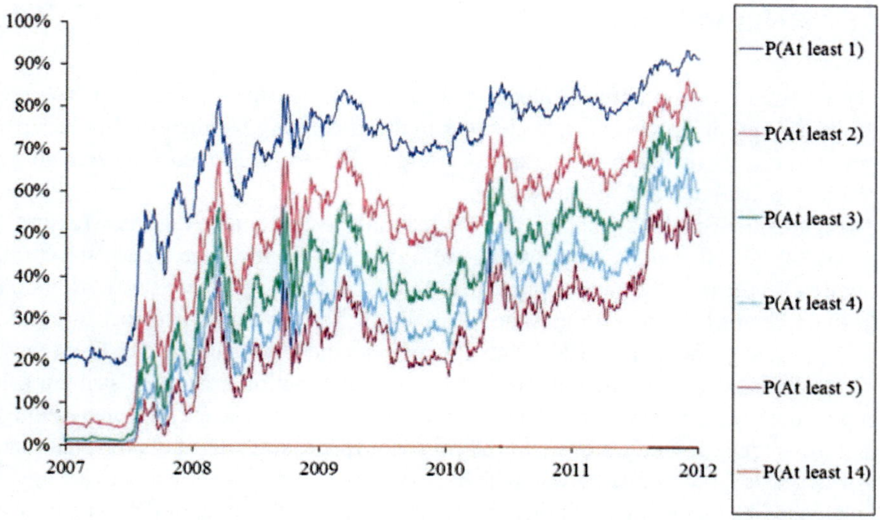

Fig. 4.11 Probability of *at least* N-1 additional banks defaulting given a particular bank defaults, involving 15 large and complex banking groups in the period 01.01.2007–31.12.2011. The median values across the cross-section of the respective probabilities are reported. The probabilities derivation incorporates empirical correlation, calculated between changes of the respective banks' 5-year CDS spreads. *Source* own calculations

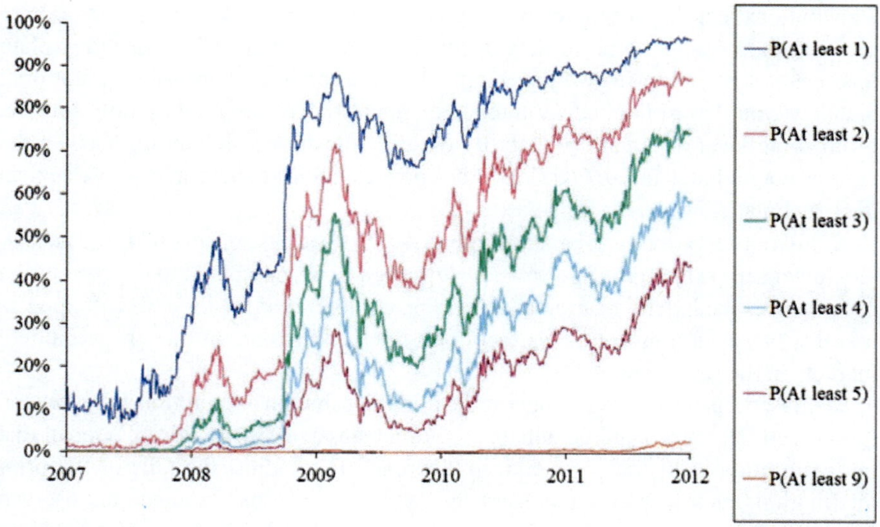

Fig. 4.12 Probability of *at least* N-1 additional sovereigns defaulting given a particular country defaults, involving 10 EA countries in the period 01.01.2007–31.12.2011. The median values across the cross-section of the respective probabilities are reported. The probabilities derivation incorporates empirical correlation, calculated between changes of the respective sovereigns' 5-year CDS spreads. *Source* own calculations

4.5 Robustness Checks

In this section, we report the results from a number of robustness checks that examine the sensitivity of our approach to changes in the underlying parameters. For brevity, only the results for the banking and sovereign SFM are presented. The remaining results are available upon request.

Figures 4.13 and 4.14 present an analysis of the Systemic Fragility Measure's sensitivity to different assumptions regarding the underlying dependence structure. Figure 4.13 presents the banking SFM results for both the cases where zero correlation (red) and empirical correlation (blue) is assumed. The former dependence structure as suggested by Segoviano (2006) and Segoviano and Goodhart (2009). In a cross-entropy context, assuming a joint standard normal distribution as a prior distribution is equivalent to assuming independence between the underlying agents. It can be shown that this independence transfers to the posterior distribution and fails to reflect the true dependence structure between the banks or sovereigns.[9] The measures are again juxtaposed to major events during the subprime and the sovereign debt crises. We document that both indicators have similar dynamics throughout the sample period and the SFM with empirical correlation is usually above the one with zero correlation. We observe a steady upward trend in the last five years, since the outbreak of the subprime crisis.

The indicator peaks at 15% for the correlation case and at 25% for the zero-correlation case. The crossing point of both versions of the indicator is surprising, as we would expect that when positive correlation between banks' asset is taken into account, banks would tend to default jointly more frequently. In order to explain this result, we need to realize that high correlation between banks means not only that they would most likely default together, but also that they might jointly survive. From some level of default probability on, this transfer of probability mass to the regions not included in Eq. (4.2) leads the correlated SFM to have a lower value than the benchmark zero-correlation case.

Similar observations can be made in Fig. 4.14 for the case of sovereigns. The usually lower series of the zero-correlation sovereign SFM, compared to the correlation case, crosses the latter one around the European Summit on 21 July 2011. It appears that this has been a milestone date both on the banking and on the sovereign debt markets in the sample period.

Our last robustness check concerns the prior distribution assumption in Segoviano (2006) and Segoviano and Goodhart (2009): a Gaussian distribution. We recall that the main purpose of the cross-entropy approach is to adjust the tails of the prior distribution, such that they represent the market consensus about the unobserved asset distribution of the system. The resulting posterior distribution is fat-tailed by construction, regardless of the choice of the prior distribution. This intuition is confirmed in Fig. 4.15 when we compare the sovereign SFM results when a normal distribution and a Student-t distribution (5 degrees of freedom) are used as priors. The cross-entropy-derived measures are almost identical in either case. Although the

[9] For details, see Sect. 3.2.2.

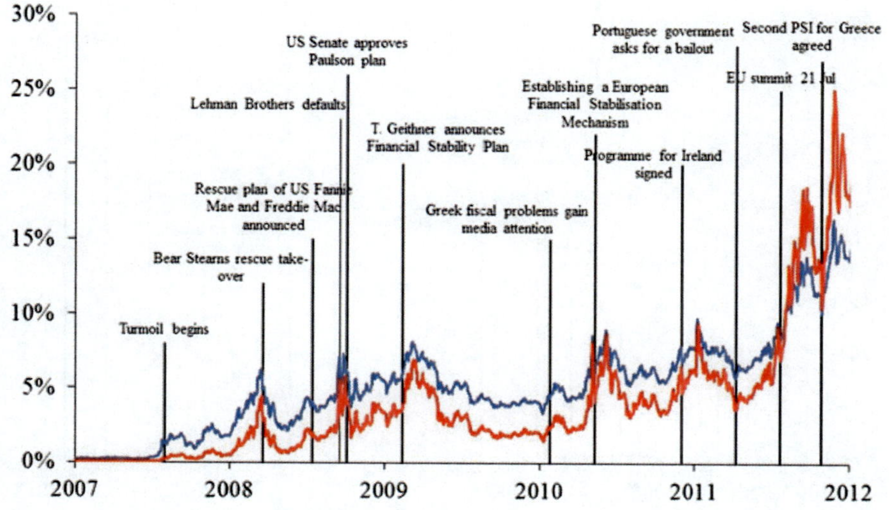

Fig. 4.13 Bank Systemic Fragility Measure (probability of at least two *banks* jointly defaulting) involving 15 large and complex banking groups in the period 01.01.2007–31.12.2011. The red line presents the results when zero correlation is assumed between banks. The blue line is calculated incorporating empirical correlation, calculated between changes of the respective banks' 5-year CDS spreads. *Source* own calculations

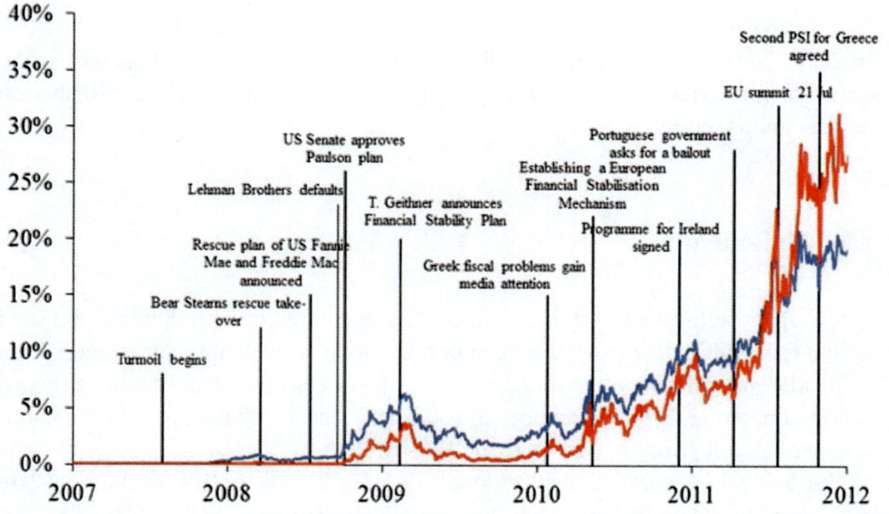

Fig. 4.14 Sovereign Systemic Fragility Measure (probability of at least two *countries* jointly defaulting) involving 10 euro area sovereigns (listed in Fig. 4.4) in the period 01.01.2007–31.12.2011. The red line presents the results when zero correlation is assumed between sovereigns. The blue line is calculated incorporating empirical correlation, calculated between changes of the respective sovereigns' 5-year CDS spreads. *Source* own calculations

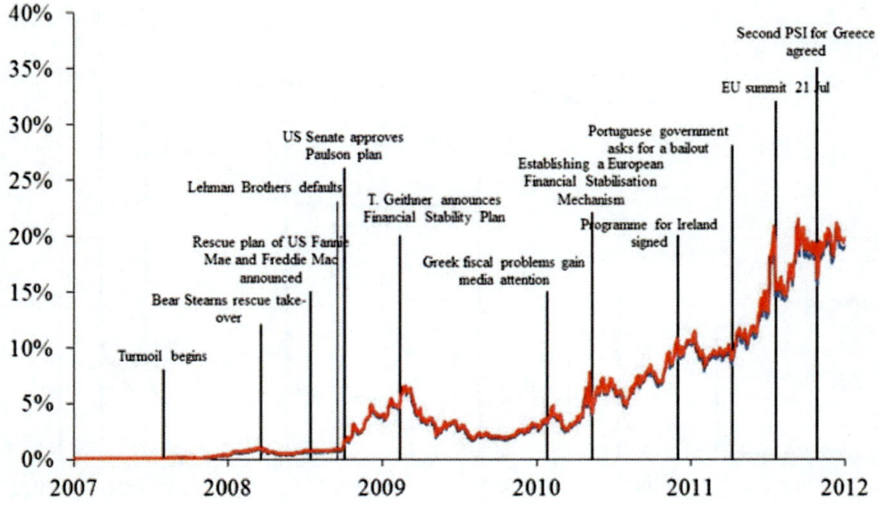

Fig. 4.15 Sovereign Systemic Fragility Measure: Normal Versus Student-T Prior Distribution. The figure plots the probability of at least two *countries* jointly defaulting for 10 euro area sovereigns (listed in Fig. 4.4) using Normal (blue line) and Student-T (red line) distributions as prior density assumptions. The measures are derived incorporating empirical correlation, calculated between changes of the respective banks' 5-year CDS spreads. Period 01.01.2007–31.12.2011. *Source* own calculations

Student-t-based measure is always above the Gaussian-based one, the latter has the same dynamics. Hence, for practical purposes, assuming a Gaussian prior distribution suffices in our setting.[10]

4.6 Conclusion

This chapter outlines a procedure for a consistent estimation of individual and joint default risk within the financial system of the euro area. We apply our methodology to calculate several measures of systemic fragility of euro area banks and sovereigns. In addition, we undertake a number of robustness checks with respect to the major parameter assumptions in our approach.

Our analysis documents that although the fragility of the EA banking system had started to deteriorate one year before Lehman Brothers' filing for bankruptcy, investors did not expect the crisis to affect EA sovereigns' solvency before September 2008. Since then, and especially after November 2009, joint sovereign default risk has outpaced the rise of systemic risk within the banking system.

[10] The relative insensitivity of the cross-entropy distribution to changes of the prior distribution assumption is also analytically shown in Segoviano (2006).

The chapter contributes to the ongoing debate on joint default risk measures and should improve our understanding of market default risk perceptions and the effects of the regulatory interventions and economic reforms needed to strengthen the euro area and prepare it for the new global financial and economic challenges. The procedure and the systemic risk measures that we propose extend the policy makers' toolkit for analysis of systemic default risk. The methodology can be easily generalized to cover not only the sample of EA banks and sovereigns, but also the global financial system.

The next chapter delves deeper into measurement of systemic risk by eliciting an important caveat that such measures should possess: They should provide an easily interpretable information about the actual *contribution* of sovereigns and banks to systemic default risk. We call this contribution "information content" and show how we could measure it in a probabilistic setting.

References

Acharya, V., Thakor, A.: The dark side of liquidity creation: leverage and systemic risk. J. Financ. Intermed. **2016**, 4–21 (2016)

Adelson, M., van Bemmelen, M., Whetten, M.: Credit Default Swap (CDS) Primer. Nomura Fixed Income Research, Nomura (2004)

Arteta, C., Hale, G.: Sovereign debt crises and credit to the private sector. J. Int. Econ. **74**(1), 53–69 (2008)

Avesani, R., Li, J., Pascual, A.: A new risk indicator and stress testing tool: a multifactor Nth-to-default CDS basket. IMF Working Papers 06/105. International Monetary Fund (2006)

EBA: 2011 EU-wide stress test aggregate report. European Banking Authority (2011)

ECB: Financial stability review. June 2010. European Central Bank (2010a)

ECB: Financial stability review. December 2010. European Central Bank (2010b)

Gorea, D., Radev, D.: The euro area sovereign debt crisis: can contagion spread from the periphery to the core? Int. Rev. Econ. Financ. **30**, 78–100 (2014)

Gramlich, D., Oet, M.: The structural fragility of financial systems: analysis and modeling implications for early warning systems. J. Risk Financ. **12**(4), 270–290 (2011)

IMF: Global financial stability report. International Monetary Fund, Washington, DC (2011)

Lehar, A.: Measuring systemic risk: a risk management approach. J. Bank. Financ. **29**(10), 2577–2603 (2005)

Lucas, A., Schwaab, B., Zhang, X.: Conditional probabilities and contagion measures for euro area sovereign default risk. J. Bus. Econ. Stat. **32**, 271–284 (2014)

Panizza, U., Sturzenegger, F., Zettelmeyer, J.: The economics and law of sovereign debt and default. J. Econ. Lit. **47**(3), 651–698 (2009)

Radev, D.: Systemic risk, sovereign default and cascade effects in the euro area. Technical report, University of Bonn (2019)

Segoviano, M.: Consistent information multivariate density optimizing methodology. FMG Discussion Papers 557. International Monetary Fund (2006)

Segoviano, M., Goodhart, C.: Banking stability measures. IMF Working Paper 09/4. International Monetary Fund (2009)

Sturzenegger, F., Zettelmeyer, J.: Haircuts: estimating investor losses in sovereign debt restructurings, 1998–2005. J. Int. Money Financ. **27**, 780–805 (2008)

Chapter 5
Systemic Risk Contributions

5.1 Motivation

Several critical issues should be considered when we attempt to measure sovereign default risk in a *systemic* context. First, we need to derive *multivariate joint* probabilities of default. As we saw in the previous chapter, multivariate joint PoDs provide more comprehensive information about the risk in the euro area than the individual PoDs of each sovereign or the bivariate PoDs of sovereign couples: In addition to the individual and bilateral country risks, they capture the complex dependence patterns and interactions among euro area countries. Second, we should be able to analyze how various scenarios involving defaults of one or several governments affect systemic risk. Hence, we need to derive *conditional* PoDs. Third, in order to investigate spillover and cascade effects among euro area sovereigns, we need to consider measures of systemic importance/contribution that examine the effect of *interdependence* on the system's vulnerability to default.

To address these important points, in this chapter we introduce a new measure of systemic risk, the change in the conditional joint probability of default ($\Delta CoJPoD$), that represents the contribution of the default of a sovereign to the overall joint default risk of the system of sovereigns.[1] More specifically, we define $\Delta CoJPoD$ as the difference between the empirical conditional probability that the euro area system[2] defaults jointly given a particular sovereign defaults, and a conditional probability, where the defaulting sovereign is assumed to be independent from the rest of the system.

$\Delta CoJPoD$ is a difference between two conditional probability measures: $CoJPoD$, derived using an *unconstrained* correlation matrix (see Table 4.5) and

[1] The overall joint default risk of the sovereign system is the probability of all sovereigns to default simultaneously. Following Lehar (2005), Adrian and Brunnermeier (2016) and Tarashev et al. (2010), we assume that the financial system is a portfolio of a finite number of European sovereigns.

[2] We constrain the euro area system to include only sovereigns.

D. Radev, *Measuring Systemic Risk*, Studies in Systems, Decision and Control 409,
https://doi.org/10.1007/978-3-030-94281-6_5

$CoJPoD'$, where we use a *constrained* correlation matrix with appropriately chosen correlations set to zero.[3] Note that the counterfactual probability $CoJPoD'$ is not an arbitrary choice. The constrained measure $CoJPoD'$ is necessary to gauge the contribution of correlation to systemic risk. We argue that focusing only on conditional probabilities of default like $CoJPoD$ may be misleading, since it does not give us information about the actual contribution of interdependence to these measures over and above the zero-correlation case and hence we cannot gauge the true extent of potential contagion effects. In the baseline definition of $\Delta CoJPoD$, we cover the direct effect of correlation. All remaining feedback effects could be captured by calculating $\Delta CoJPoD$ as a Shapley value by averaging across the systemic risk contributions to default risk of all possible coalitions that include a particular sovereign or a group of sovereigns.

The difference between these two conditional probabilities is important, because it accounts for the heterogeneity in interconnectedness and interdependence among countries. Some countries are more interconnected than others, for instance, through trade relationships or financial integration. Let us assume that EA sovereign A is more interconnected with EA sovereign B than with EA sovereign C. Then, if sovereign A defaults, the probability that sovereign B will default should be higher than the probability that sovereign C defaults. To fully capture the heterogeneity in the impact of sovereign A's default on the PoDs of sovereigns B and C, we need to compare them to the counterfactual cases when sovereigns B and C, respectively, are not interconnected to sovereign A at all. Therefore, $\Delta CoJPoD$ captures the *additional* default vulnerability of sovereigns B and C that arises from a default of sovereign A.

More generally, $\Delta CoJPoD$ represents the difference between a potentially informative conditional probability and an uninformative (counterfactual) conditional probability. The informative part provides additional information about the vulnerability of the euro area system, conditional on the default of a sovereign. This additional information needs to be disentangled from the uninformative portion, where the default of the sovereign has no impact on the system. Hence, $\Delta CoJPoD$ focuses on the additional information that conditional default probabilities provide and therefore represents the *information content* in a sovereign default. In Sect. 5.5.1, we show that focusing solely on conditional probabilities without considering the actual additional information that they provide could lead to *spurious conclusions* and a bias in the estimation of the actual systemic risk.

As in the previous chapter, we apply a three-step procedure to estimate sovereign systemic risk contributions. First, we derive *individual* PoDs from each sovereign's credit default swap spread series,[4] using a CDS-bootstrapping procedure that follows Hull and White (2000). Second, since *joint* default risk is not traded, we need to impose a flexible structure on the interdependence between the individual sovereigns

---

[3] See Table 5.2 for an example of the correlation matrix for the counterfactual that we use to derive the individual contribution of Greek government default to sovereign systemic risk.

[4] The literature on credit default swap pricing uses the terms "CDS spreads", "CDS prices" and "CDS premia" interchangeably. For the remainder of the chapter, we will use these three terms as synonyms.

under investigation. We employ a minimum cross-entropy copula-based approach to approximate the unknown multivariate distribution of EA sovereign. Third, we calculate the new systemic risk measure, the change in the Conditional Joint Probability of Default ($\Delta CoJPoD$) using the derived multivariate density and investigate its properties. In the subsequent empirical analysis, we use our measure to assess the systemic importance of euro area sovereigns and to analyze cascade effects among these sovereigns.

Our results show that *joint* sovereign distress risk has increased since the end of 2009, parallel to a decoupling of investors' perceptions about *individual* sovereign default risk. We find that Germany and the Netherlands have the highest systemic importance in the euro area, while the effect of a default of Greece is marginal at best. The latter effect is mainly due to the low correlation of Greek sovereign assets with the rest of the euro area. The size of the bias-correction varies from about 2.5% for Germany to more than 80% for Greece, which underlines the importance of the accurate estimation of the information content in a sovereign default, especially for distressed countries.

To provide an example for the flexibility of $\Delta CoJPoD$ for policymaker purposes, we apply it to analyze the potential cascade effects within the euro area, concentrating on a particular scenario: We investigate how the perceptions about a default of a relatively small EA sovereign (Portugal) affect the default expectations about another small sovereign (Ireland), and how the expectations about their *joint* default impact the default perceptions about a larger sovereign (Spain). This scenario was of particular interest during the Sovereign debt crisis, since the problems of small countries like Greece, Ireland and Portugal affected the perceptions about the default risk of larger, troubled countries like Spain and Italy. We find that a hypothetical default of Portugal would have increased the probability of default of Ireland by almost 60 *percentage points* (from the fairly low unconditional probability of 6% to the substantial conditional probability of 65%) in early 2009. Therefore, our measure would have sent an early warning signal long before these two countries had to be bailed out in 2010 and 2011. Further, we find a high probability of distress spillover between the small sovereigns and Spain: a hypothetical joint default of the latter two countries would have increased the default risk of Spain on average by 40 percentage points from the outbreak of the subprime crisis throughout the European Sovereign debt crisis. Eventually, Spain applied for bailout funds as well. Section 5.3.4 shows that although the individual systemic risk contribution of Spain has subsided in recent times, the risk from possible cascade effects from Ireland and Portugal remains at a very high level. Therefore, we argue that possible default cascade effects of reasonable size should be taken into account in future political decision-making, even if the source of the cascade may be considered relatively immaterial in size.

The new measure is an important contribution to the literature, because it represents a bias-corrected perception about the increase in systemic risk that arises from a default of a sovereign. The bias stems from ignoring the actual interdependence between the participants in the financial system when we analyze only conditional probabilities. We argue that neither the unconstrained $CoJPoD$, nor the constrained $CoJPoD'$ can provide a full picture of the contagion potential if used on their own and we need to take the difference, in order to arrive at a measure that shows the contribution of correlation beyond the zero-correlation case. In case that a sovereign has a relatively low interdependence with the rest of the sovereigns in the system, the contribution on the systemic risk of the euro area conditional on its default should be limited. To illustrate this point, consider the case of Greece. If we have a relatively low average correlation, as Greece does (see Table 4.5), the "bias-correction" is relatively large (about 50%), because the true contagion effect due to interdependence is low relative to the zero-correlation case. As our results show, low correlation has led to a very low systemic contribution of Greek default to the risk of the euro area. Our findings are confirmed by the small international impact of the Greek sovereign debt restructuring in early 2012, despite the imposed large haircuts between 50 and 70%.[5] When we only analyze conditional probabilities and not $\Delta CoJPoD$, we may still observe non-zero dynamics conditional on Greek default, which will be misleading for policy analysis. In contrast, by subtracting the appropriate counterfactual, $\Delta CoJPoD$ correctly shows that the contribution of Greece is minimal. This point is further elaborated upon in Sects. 5.3.4 and 5.5.1. These pitfalls of conditional default probabilities have not yet been analyzed in the financial stability literature.

The *direct* contribution on systemic risk of a default of a single sovereign does not, of course, represent the complete effect of the default on systemic risk. For instance, the effect of a Greek default may affect the system via an increase of the riskiness of Portugal or Italy. These relationships are captured to some extent by the correlations in the unconstrained $CoJPoD$, but as an extension, we provide scenarios with cascade effects, where we constrain two or more countries, depending on the effect that we need to calculate. To assess the full effect of a country's default on systemic risk, the measure can be extended to a full Shapley-value type measure that is an average of all possible permutations of coalitions of defaulting sovereigns that include the sovereign under investigation.[6] The logic and benefit of $\Delta CoJPoD$ remains unchanged and we argue that it is a significant contribution to the literature on systemic risk and minimum cross-entropy probability measures.[7]

$\Delta CoJPoD$ seems to be most related to the "spillover component" by Lucas et al. (2014), who analyze the difference of the probability of default of Portugal, conditional on the default of Greece and the probability of default of Portugal conditional on Greece not defaulting. Apart from being a bivariate measure (hence, not a systemic measure), the main difference of the "spillover component" to the multivariate $\Delta CoJPoD$ is in the definition of the counterfactual conditional probability, and therefore the measure cannot be interpreted as the *information content* of sovereign default. Moreover, it can be shown that the "spillover component" reflects the *average observed effect* of a default. However, regulators can either observe a default or a non-default, and never both counterfactuals at the same time, and therefore the calculation of the effect suffers from a *selection bias* (see, Angrist and Pischke 2009,

[5] If the average correlation is high, we have a smaller correction, but the bias still exists.

[6] For a full definition of the Shapley value, see Tarashev et al. (2010).

[7] The reason why we focus on the most simple case first is that systemic risk measures need to be understandable to policymakers that make many parallel decisions in a short period of time.

p. 22), since a sovereign can self-select to default or not. This effect is therefore of limited interest for causal analysis. In Sect. 5.2, we show that $\Delta CoJPoD$ reflects the *average treatment effect on the treated* and provides important additional information to regulators and policy-makers. Section 5.5.2 contains a more in-depth discussion regarding the "spillover component" measure, as well as the remaining measures related to the $\Delta CoJPoD$.

The chapter is organized as follows. In Sect. 5.2, we introduce the $\Delta CoJPoD$ measure and propose a procedure to derive it. Section 5.3 discusses how $\Delta CoJPoD$ can be used to measure systemic contributions and cascade effects among sovereigns. Section 5.5 compares the new measure with other measures of systemic risk. Section 5.6 concludes by providing the main policy implications of our analysis and suggesting possible applications of $\Delta CoJPoD$ in regulatory decision-making.

5.2 Conditional Joint Probability of Default

5.2.1 *Definition*

At the basis of our conditional probability measure is the joint (unconditional) probability of default (*JPoD*) of the system, which can be interpreted as the *system's overall vulnerability to default events*.

Let the system be described by an n-dimensional joint distribution, $P(x_1, x_2, \ldots, x_n)$, with density $p(x_1, x_2, \ldots, x_n)$, where $x_1, x_2, \ldots, x_n$ are the logarithmic assets of the respective sovereign $X_1, X_2, \ldots, X_n$. Following Segoviano and Goodhart (2009), we then define the Joint Probability of Default (*JPoD*) as:

$$JPoD_{x_1,x_2,\ldots,x_n} = JPoD_{system} = \int_{\overline{\mathbf{x}}_1}^{+\infty}\int_{\overline{\mathbf{x}}_2}^{+\infty}\cdots\int_{\overline{\mathbf{x}}_n}^{+\infty} p(x_1, x_2, \ldots, x_n)dx_1, dx_2 \ldots dx_n \tag{5.1}$$

where $\overline{\mathbf{x}}_1$, $\overline{\mathbf{x}}_2$, ..., $\overline{\mathbf{x}}_n$ are the individual default thresholds[8] of the respective sovereigns.[9]

---

[8] The default thresholds are defined in the sense of the classical structural model by Merton (1974): an entity defaults if its assets drop below a specific value. Note that as in Segoviano (2006) and Segoviano and Goodhart (2009), the default region is in the right tail of the distribution. This does not affect our results, due to the assumption of a symmetrical prior distribution, but simplifies our estimation procedure.

[9] As in Segoviano and Goodhart (2009), we define the default threshold to be:

$$\overline{\mathbf{x}}_k = \Phi^{-1}(1 - \overline{PoD^k}),$$

where $\Phi^{-1}(\cdot)$ is the inverse of the cumulative distribution function of a standard normal distribution and $\overline{PoD^k}$ is the through-the-period average of the CDS-derived individual probabilities of default of sovereign k.

Then, applying Bayes' rule, we derive the Conditional Joint Probability of Default of the system of n sovereigns, conditional on sovereign k defaulting:

$$\begin{aligned} CoJPoD_{system_{-k}|x_k>\overline{\mathbf{x}}_k} &= JPoD_{x_1,x_2,\ldots,x_{k-1},x_{k+1},\ldots,x_n|x_k>\overline{\mathbf{x}}_k} \\ &= \frac{JPoD_{x_1,x_2,\ldots,x_n}}{PoD^k} \\ &= \frac{JPoD_{system}}{PoD^k}, \end{aligned} \tag{5.2}$$

where PoD^k is the individual default probability of sovereign k.[10] Therefore, the conditional probability of default is a ratio between the general vulnerability of the system and the individual default vulnerability of a sovereign.

Note that, by definition, Bayes' rule yields the conditional probability of the *remaining* (non-defaulting) sovereigns in the system. It does not, however, convey information about what their probability would have been without the shock due to sovereign k's default. We seek to compare the conditional probability of default of the surviving sovereigns with their *unconditional* probability of default, which may also be considered their general vulnerability during "tranquil" times. Therefore, to calculate the contribution of sovereign k's default on the system's default risk, we subtract from $CoJPoD_{system_{-k}|x_k>\overline{\mathbf{x}}_k}$ the unconditional *JPoD* of the system constituents *excluding* the sovereign in question. Our $\Delta CoJPoD$ measure is then

$$\Delta CoJPoD_{system_{-k}|x_k>\overline{\mathbf{x}}_k} = CoJPoD_{system_{-k}|x_k>\overline{\mathbf{x}}_k} - JPoD_{system_{-k}}. \tag{5.3}$$

In essence, we compare the risk of the system when sovereign k is included and defaults, to a situation, in which sovereign k is excluded, or otherwise said—independent from the system. So defined, $\Delta CoJPoD$ is the probabilistic alternative to the CoVaR (Adrian and Brunnermeier 2016).

Next, let us define $JPoD'_{system}$ as the joint probability of default of the system if sovereign k is independent from the rest of the system, all other things equal. The prime indicates that the only difference between $JPoD_{system}$ and $JPoD'_{system}$ is that in $JPoD'_{system}$ we assume independence between sovereign k and every other sovereign in the system. Both probability measures are identical in all other aspects of their underlying dependence structure. Applying Bayes' rule, we can reformulate $JPoD'_{system}$ as

[10] The individual probability of default of sovereign k is defined as

$$\int_{-\infty}^{+\infty}\int_{-\infty}^{+\infty}\cdots\int_{-\infty}^{+\infty} p(x_1, x_2, \ldots, x_n)\mathbf{I}_{[\overline{\mathbf{x}}_1,\infty)}dx_1\cdots dx_{n-1}dx_n = PoD^k,$$

where $\mathbf{I}_{[\overline{\mathbf{x}}_1,\infty)}$ is an indicator variable that takes the value of 1 if the latent asset process of sovereign k crosses the sovereign-specific default threshold and 0 otherwise. In practice, it is estimated empirically from CDS spreads using a bootstrapping procedure. The estimation procedure is outlined in Sect. 3.1.

$$
\begin{aligned}
JPoD'_{system} &= JPoD'_{x_1,x_2,\ldots,x_{k-1},x_k,x_{k+1},\ldots,x_n} \\
&= JPoD'_{x_1,x_2,\ldots,x_{k-1},x_{k+1},\ldots,x_n|x_k>\bar{\mathbf{x}}_k} \cdot PoD^k \\
&= JPoD'_{x_1,x_2,\ldots,x_{k-1},x_{k+1},\ldots,x_n} \cdot PoD^k \\
&= JPoD_{system_{-k}} \cdot PoD^k .
\end{aligned} \tag{5.4}
$$

Then, $JPoD_{system_{-k}}$ can also be represented in the following way:

$$
\begin{aligned}
JPoD_{system_{-k}} &= \frac{JPoD'_{x_1,x_2,\ldots,x_n}}{PoD^k} \\
&= CoJPoD'_{system_{-k}|x_k>\bar{\mathbf{x}}_k},
\end{aligned} \tag{5.5}
$$

where $CoJPoD'_{system_{-k}|x_k>\bar{\mathbf{x}}_k}$ is the conditional counterpart of $JPoD'_{system}$ with respect to sovereign k. Thus, our systemic risk contribution from Eq. (5.3), $\Delta CoJPoD_{system_{-k}|x_k>\bar{\mathbf{x}}_k}$, transforms to

$$
\Delta CoJPoD_{system_{-k}|x_k>\bar{\mathbf{x}}_k} = CoJPoD_{system_{-k}|x_k>\bar{\mathbf{x}}_k} - CoJPoD'_{system_{-k}|x_k>\bar{\mathbf{x}}_k}, \tag{5.6}
$$

where we use the **unconstrained** correlation matrix in Table 4.5 to derive $CoJPoD_{system_{-k}|x_k>\bar{\mathbf{x}}_k}$ and the **constrained** correlation matrix in Table 5.2 to derive $CoJPoD'_{system_{-k}|x_k>\bar{\mathbf{x}}_k}$.[11]

The measure can then be viewed as the difference between the (potentially) *informative* conditional probability $CoJPoD_{system_{-k}}$ and the *uninformative* (due to the independence assumption) conditional probability $CoJPoD'_{system_{-k}}$. Thus, $\Delta CoJPoD_{system_{-k}}$ measures the *information content* of a default of sovereign k, which could also be viewed as the contribution to the systemic default risk due to the system's interdependence with sovereign k.

In Eq. (5.6), we can notice that $\Delta CoJPoD$ can also be interpreted as a generalization of the Average Treatment Effect on the Treated (ATET), which is of significant importance for causal analysis. Ideally, we would be interested in the ATET, as well as the effect on the non-treated and on the overall population. However, due to a selection bias (Angrist and Pischke 2009, p. 22), the latter two effects are difficult to identify and therefore of limited application in economics and finance.[12]

---

[11] Find Tables 4.5 and 5.2 in Sect. 5.3.3. In the particular example in Table 5.2, we use Greece as the defaulting sovereign.

[12] The observed difference in average outcomes can be summarized as (Angrist and Pischke 2009):

$$
E[Y_i|D_i=1] - E[Y_i|D_i=0] = E[Y_{1i}|D_i=1] - E[Y_{0i}|D_i=1] + E[Y_{1i}|D_i=1] - E[Y_{0i}|D_i=0], \tag{5.7}
$$

where D_i is a dummy equal to 1, in case the object (the sovereign system) is treated (a sovereign k defaults) and 0 otherwise; Y_i is the observed outcome, which is further equal to Y_{1i} if the sovereign system experiences a default of sovereign k and Y_{0i} in case that it does not.

Examining Eq. (5.6) more carefully and realizing that the CDS-derived probabilities of default are market expectations about the solvency of sovereign, we can translate the $\Delta CoJPoD$ measure to the term

We should note that in our approach interdependence does not necessary mean direct interconnectedness. Sovereigns with no or minor direct financial and trade linkages could still be interdependent if the markets perceive them to be similar in any way. This perceived similarity causes a comovement of their individual CDS series, which reflects a comovement of market perceptions about the default of the individual sovereigns. Therefore, analogously to the CoVaR, our measure does not reflect a causal relationship. Nevertheless, we could witness a directionality in our conditional indicator: the *CoJPoD* of sovereign A defaulting given that sovereign B defaults could be different from the *CoJPoD* of sovereign B defaulting given sovereign A defaults.

5.2.2 Estimation Procedure

There are numerous ways to calculate the individual and joint probabilities of default to derive $\Delta CoJPoD_{system_{-k}|x_k}$. To calculate individual probabilities of default (PoD), we choose a bootstrapping procedure that incorporates all available CDS contracts of a sovereign up to a 5-year horizon. Then, we transform the *individual PoDs* into *multivariate JPoDs* using a copula-based minimum cross-entropy procedure.

We should point out that $\Delta CoJPoD$ is a general concept that is independent from the estimation procedure of choice. It can be calculated, for instance by using a multivariate t-distribution as in Lucas et al. (2014, 2017), and such a method will employ variable thresholds, as opposed to the fixed ones in the minimum cross-entropy approach. Unreported robustness checks showed minor quantitative differences between both calculation methods, while the qualitative conclusions remained the same.

A commonly overlooked property of the usual applications of the minimum cross-entropy approach in banking literature (see, e.g., Segoviano and Goodhart 2009) is that if independence is assumed for the prior distribution (e.g., by assuming a zero-cross-correlation structure for the prior distribution, as in Segoviano 2006), this transfers to the posterior distribution as well. Section 3.2.2 provides a multivariate proof of this caveat when multivariate joint normal distribution is assumed as a prior. In general, zero correlation does not imply independence and simple analytical examples are readily available. However, if zero correlation is assumed for a joint normal asset distribution, the resulting joint probabilities of default are a product of the individual (marginal) probabilities of default. Hence, any systemic probability measure that conditions on particular entities defaulting will be equal to the product of the PoDs of the remaining entities. This is equivalent to saying that by conditioning on a default of an entity in a zero-cross-correlation setting, we do not obtain

---

$$E\left[Y_{1i}|D_i = 1\right] - E\left[Y_{0i}|D_i = 1\right], \tag{5.8}$$

which is the Average Treatment Effect on the Treated, with $CoJPoD'$ being the counterfactual of the system, when it is not affected by the default of sovereign k, and hence equal to $E\left[Y_{0i}|D_i = 1\right]$. The second difference on the right-hand side of Eq. (5.7) is the selection bias.

additional information about the default of the remaining entities, apart from the information already contained in their individual probabilities of default. That fact exactly complies with the probabilistic definition of independence.

A failure to take this observation into account has affected previous studies. In a recent work, Peña and Rodriguez-Moreno (2013) compare the predictions of several systemic risk models, including the entropy-derived Banking Stability Index (BSI), but assume a zero-correlation structure for the prior distribution. If this assumption fails, which is likely to be the case for bank assets, there will be a significant underestimation of the joint default risk between the considered entities.

Additionally, due to the independence of the posterior distribution, any conditional measures derived using it will be identical to their unconditional counterparts. The latter fact has a pronounced effect on our $\Delta CoJPoD$ measure, as it is exactly the difference between the conditional JPoD and its unconditional alternative. It can be shown that this measure will be exactly 0 at any point of time, despite any dynamics in the individual PoDs. Empirical evidence for this analytical result is provided in Sects. 5.3.3 and 5.5.1.

Since the initial correlation structure assumption is crucial for the minimum cross-entropy approach, we rely on market data to explicitly allow it to differ from the identity matrix. This distress correlation structure is proxied by the empirical correlation between changes in the 5-year CDS spreads of the sovereigns in our sample.

5.3 Euro Area Sovereign Default Risk

Our empirical analysis is organized as follows. First, we investigate the default risk contributions among 10 euro area sovereigns. With sovereign default still a major concern in the EA, it is important to analyze the dynamics of our systemic risk measure and identify possible trends, as well as major regulatory interventions and their effects. Then, we focus on the possible cascade effects among EA sovereigns.

5.3.1 Data and Descriptive Statistics

We estimate marginal probabilities of default using CDS premia for contracts with maturities from 1 to 5 years for the period 01/01/2008 and 04/27/2015. The employed procedure (for details, see Hull and White 2000 and Sect. 3.1) requires as additional inputs refinancing interest rates, which we choose to be the AAA euro area government bond yields for maturities from 1 to 5 years. The CDS spreads and the government bond yields are at daily frequency, which is also the frequency of the resulting probabilities of default. Our analysis covers 10 euro area (EA) sovereigns (Austria, Belgium, France, Germany, Greece, Ireland, Italy, the Netherlands, Portugal and Spain). In order to arrive at comparable CDS-derived probabilities of default, all components in the calculation should be under a common currency measure. For

Table 5.1 Descriptive statistics of the 5-year CDS spread series of Austria (AT), Belgium (BE), France (FR), Germany (GE), Greece (GR), Ireland (IE), Italy (IT), the Netherlands (NL), Portugal (PT), Spain (ES). The data are in basis points. Period: 01/01/2008–04/27/2015. Number of observations: 1910

	AT	BE	FR	GE	GR
Minimum	6.00	10.60	6.70	4.50	22.10
Mean	72.53	100.09	73.80	38.96	2578.40
Maximum	273.00	403.01	247.08	120.59	25960.76
Standard deviation	54.21	79.36	52.65	25.15	4471.47
Number of observations	1910	1910	1910	1910	1910
	IE	IT	NL	PT	ES
Minimum	13.30	20.30	6.50	17.70	17.10
Mean	273.85	195.71	49.30	382.50	197.52
Maximum	1286.91	595.68	136.21	1762.10	635.58
Standard deviation	252.28	131.60	30.43	362.31	138.46
Number of observations	1910	1910	1910	1910	1910

many of the sovereigns both euro and US-dollar-denominated CDS contracts are traded. In an unreported analysis, we came to the conclusion that the difference in the absolute levels of the series cannot be explained solely by the exchange rate dynamics. As CDS contracts are usually traded over the counter, it is difficult to find information on the exact traded volumes of each type. After consulting professionals, we were assured that in the case of sovereigns, the US-dollar-denominated contracts are more liquid. For this reason, where available, these were chosen in our analysis and the data was transformed using euro-dollar exchange rates, downloaded from Bloomberg. For consistency, the CDS contracts are denominated in Euros. The sources for sovereign CDS spreads and government bond yields are the Intercontinental Exchange Competition and Markets Authority (ICE CMA) and Datastream, respectively. The exchange rate quotes are downloaded from Bloomberg.

Table 5.1 presents the descriptive statistics of the 5-year CDS spreads of the ten countries in our sample. The average 5-year CDS spread in the cross-section ranges from 39 basis points for Germany to 2578 basis points for Greece. We also notice a substantial increase of CDS premia even for the safest country, Germany, from 4.5 to 121 basis points. However, this does not compare to the dynamics of the price for protection against the default of Greece, which starts at 22 basis points at the beginning of the period and reaches almost 26000 basis points on March 8, 2012 when the restructuring of the Greek sovereign debt (the above-mentioned PSI) was announced. We also notice that the safest countries, Germany and the Netherlands, exhibit the lowest volatility in the price for protection against default.

5.3.2 *Marginal Probabilities of Default*

Figure 5.1 depicts the CDS-implied annualized probabilities of default for the 10 euro area sovereigns in our analysis. We observe very similar values in the beginning of our sample period, pointing towards investors' confidence in the individual EA members' ability to service their debt. We observe a peak in the individual PoDs during the global recession after Lehman Brother's collapse, but the individual default risk gradually falls throughout 2009. A major decoupling occurs in November 2009, after the announcement by the newly-elected Greek government that the previously reported data on the government deficit was strongly misleading. The following divergence of market expectations about individual sovereign default risk may be due not only to doubts about the *individual* governments' ability to service their debt, but also about the potential of the euro area as a whole to support its members in need. We also notice that the PoD level of Greece rises until the end of 2013, as previously noted, while the default risk perceptions with regard to the remaining distressed EA countries—Ireland, Italy, Portugal and Spain—stabilize in the first half of 2012. Overall, it can be noted that the dynamics of the Greek sovereign PoD diverges substantially from the remainder of the EA system.

Figure 5.2 provides a graphical representation of the sensitivity of the bootstrapped PoDs (for Greece in this case) to changes in the assumed recovery rate. The figure shows that the probability of default rises monotonically with higher recovery rates. Therefore, if we were to assume a higher recovery rate in our analysis, this would

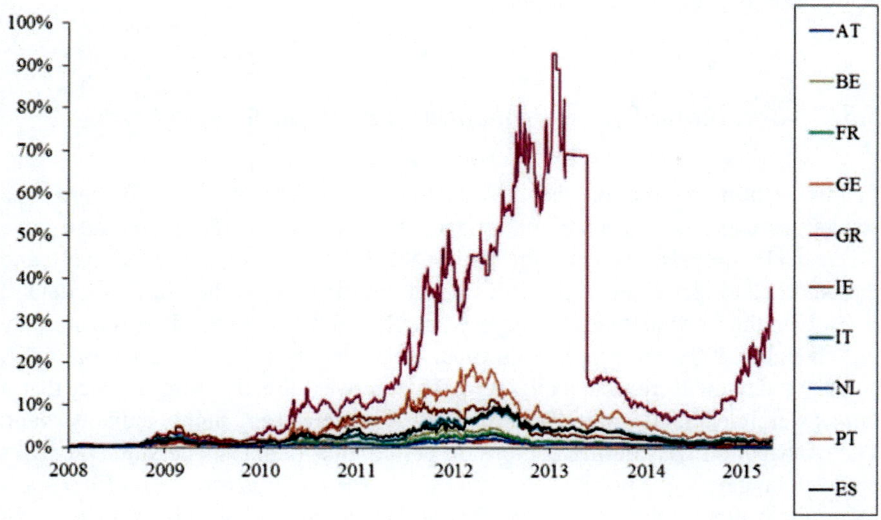

Fig. 5.1 5-year annualized CDS-implied bootstrapped probabilities of default for 10 sovereigns: Austria (AT), Belgium (BE), France (FR), Germany (GE), Greece (GR), Ireland (IE), Italy (IT), the Netherlands (NL), Portugal (PT), Spain (ES). Euro-denominated CDS spreads are used. Period: 01/01/2008–04/27/2015. *Source* own calculations

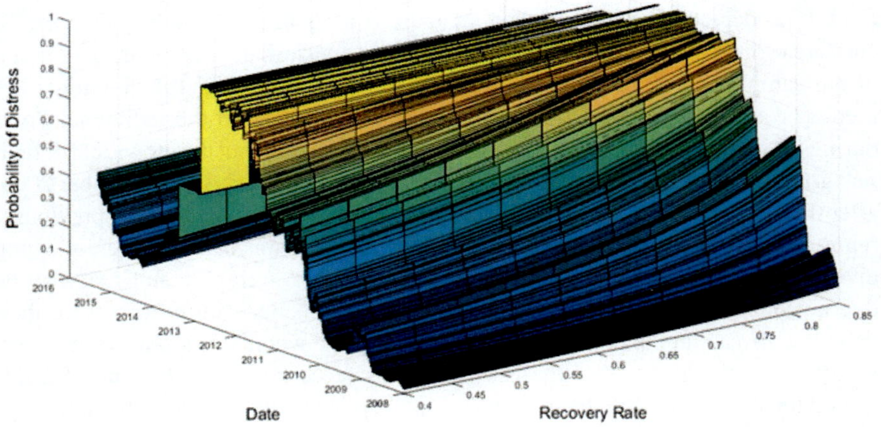

Fig. 5.2 5-year annualized probability of default of Greece with different recovery rates. The probabilities are derived using a survival probability model (CDS bootstrapping), based on Hull and White (2000). Period 01/01/2008–04/27/2015. *Source* own calculations

translate into higher PoDs and, consequently, into a higher CoJPoD. We assume a relatively conservative recovery rate of 40%, since it is in line with the haircuts and recovery rates recorded for Greece—the only sovereign in the euro area that had to restructure its public debt in the sample period.

5.3.3 Conditional Joint Probabilities of Default

In this section, we present the $\Delta CoJPoD$ results for our set of 10 euro area sovereigns using the correlation matrix of the bilateral changes in the respective 5-year CDS spreads, presented in Tables 5.2 and 5.3. In Fig. 5.3, Germany and the Netherlands have the highest perceived contribution to the euro area default risk, given their own default. We observe that before Lehman Brothers' bankruptcy in September 2008, the perceptions about the systemic risk contribution of an EA country's default were practically zero. This derives directly from the fact that a *joint* sovereign default within the euro area was perceived as a highly unlikely event. The contribution rises during the turmoil period after Lehman's default, and peaks between January and April 2009, gradually subsiding afterwards. $\Delta CoJPoD$ starts rising again after the announcement of the budget problems of the Greek government in November 2009 and peaks at nearly 8 percentage points for Germany at the end of November 2011.

A more elaborate interpretation of the $\Delta CoJPoD$ is that its first part, the *CoJPoD*, reflects the relative dynamics of systemic fragility ($JPoD_{system}$) and the individual

Table 5.2 Correlation structure among 10 sovereigns for calculating $CoJPoD$: Austria (AT), Belgium (BE), France (FR), Germany (GE), Greece (GR), Ireland (IE), Italy (IT), the Netherlands (NL), Portugal (PT), Spain (ES). Period: 01/01/2008–04/27/2015. The correlations are calculated between changes in the 5-year CDS spreads of the sovereigns in the respective column and row

	AT	BE	FR	GE	GR	IE	IT	NL	PT	ES
AT	1.00	0.70	0.73	0.75	0.13	0.54	0.66	0.79	0.46	0.62
BE		1.00	0.82	0.74	0.16	0.69	0.83	0.72	0.65	0.81
FR			1.00	0.82	0.22	0.63	0.81	0.73	0.60	0.76
GE				1.00	0.19	0.59	0.72	0.75	0.56	0.69
GR					1.00	0.19	0.21	0.16	0.17	0.17
IE						1.00	0.71	0.55	0.77	0.74
IT							1.00	0.68	0.71	0.90
NL								1.00	0.49	0.63
PT									1.00	0.73
ES										1.00

Table 5.3 Correlation structure among 10 sovereigns for calculating $CoJPoD'$: Austria (AT), Belgium (BE), France (FR), Germany (GE), Greece (GR), Ireland (IE), Italy (IT), the Netherlands (NL), Portugal (PT), Spain (ES). Period: 01/01/2008–04/27/2015. The correlations are calculated between changes in the 5-year CDS spreads of the sovereigns in the respective column and row. In this example, Greece is assumed to be uncorrelated with the rest of the system, in order to calculate the respective $CoJPoD'$, given Greece defaults

	AT	BE	FR	GE	GR	IE	IT	NL	PT	ES
AT	1.00	0.70	0.73	0.75	**0.00**	0.54	0.66	0.79	0.46	0.62
BE		1.00	0.82	0.74	**0.00**	0.69	0.83	0.72	0.65	0.81
FR			1.00	0.82	**0.00**	0.63	0.81	0.73	0.60	0.76
GE				1.00	**0.00**	0.59	0.72	0.75	0.56	0.69
GR					**1.00**	**0.00**	**0.00**	**0.00**	**0.00**	**0.00**
IE						1.00	0.71	0.55	0.77	0.74
IT							1.00	0.68	0.71	0.90
NL								1.00	0.49	0.63
PT									1.00	0.73
ES										1.00

default risk of a sovereign (PoD_k). For the case of Germany, although its individual risk increases slightly but steadily throughout the sample period, the overall EA systemic fragility rises at a faster (or falls at a slower) pace. At the other end of the spectrum is Greece, where the individual risk dynamics outpace those of the system, in terms of both growth and magnitude, resulting in a lower risk contribution due to interdependence/correlation. A positive result for the risk contribution $\Delta CoJPoD$ means that, due to the interconnectedness of the respective sovereign with the rest of the euro area, the fragility of the system increases more than if the country default is

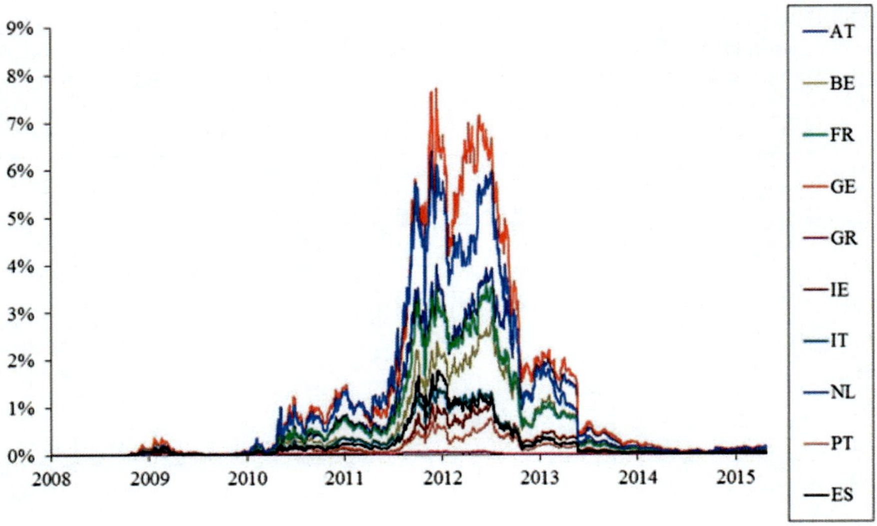

Fig. 5.3 Change in the conditional joint probabilities of default ($\Delta CoJPoD$) involving 10 sovereigns for the period 01/01/2008–04/27/2015. Non-zero correlation structure is used in defining the prior distribution. The correlation matrix is presented in Table 4.5. The correlation structure calculation is explained in the text and in Table 4.5. For an explanation of the abbreviations, see Fig. 4.4. $\Delta CoJPoD$ is the difference between the respective series presented in Eq. (5.6)

an independent event. Overall, the results for $\Delta CoJPoD$ imply that this difference is much higher for Germany than for the default of any other country. The effect of a default of Greece is neutralized by the low correlation with the remainder of the system, which is visible in Table 5.2. Hence, the *information content* in a Greek default regarding the fragility of the overall EA sovereign system is very low. This is confirmed by the events around the Greek sovereign debt restructuring in March 2012, when the risk of the other nine countries in the sample did not increase, despite the fact that their banks agreed to significant haircuts on their Greek sovereign debt holdings.

There is a second effect contributing to the final results, apart from pure dependence, namely the level effect of the systemic and individual default risks. As the individual level of default risk of Greece is high compared to the systemic default risk level, *CoJPoD* will be low, leading to low results for $\Delta CoJPoD$, as well. The benefit of our measure is that it takes into account the interaction of both effects when evaluating the overall impact of interdependence on systemic risk.

The effects of low correlation are taken to their extreme in Fig. 5.4, where we present the results for $\Delta CoJPoD$ if the countries are assumed to be independent. As argued in Sect. 3.2, in that case the contribution to systemic risk of any of the sovereigns is by definition not different from 0, due to the fact that under independence, the conditional and unconditional JPoDs are the same.[13] The visual

[13] The spikes in certain periods are due to minor rounding errors.

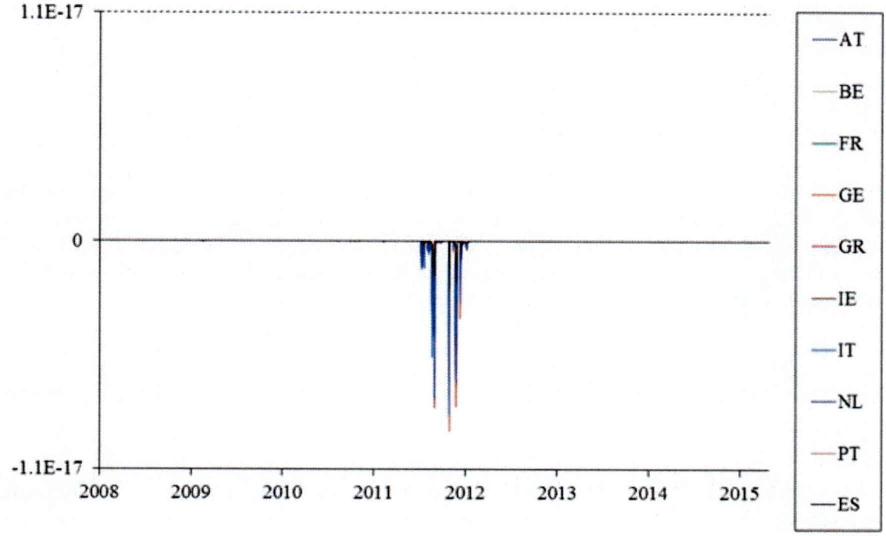

Fig. 5.4 Change in the conditional joint probabilities of default ($\Delta CoJPoD$) involving 10 sovereigns for the period 01/01/2008–04/27/2015. Zero correlation between all sovereigns involved is assumed in defining our prior distribution. $\Delta CoJPoD$ is the difference between zero-correlation analogues of the respective series presented in Eq. (5.6). This plot confirms the analytical result that if independence is assumed for the prior distribution, it transfers to the CIMDO posterior distribution, yielding $\Delta CoJPoD = 0$. For an explanation of the abbreviations, see Fig. 4.4

of Fig. 5.4 confirms that if we assume independence among sovereigns, this will lead to misleadingly low contributions to systemic risk.

Revisiting Fig. 5.3, we notice that the ordering of the conditional joint default contributions of the euro area countries is a mirror image of their marginal probabilities of default in Fig. 4.4, although the dynamics in both cases is quite different. This stems from the definition of the Bayes' rule in Eq. (5.2): given the same $JPoD_{system}$, $CoJPoD$ is the inverse of the marginal probabilities of default. The interpretation of $CoJPoD$ and $\Delta CoJPoD$ becomes less trivial when we investigate cascade effects, that is, the defaults of several countries in a sequence, which is the more interesting application of the approach and the subject of the next section.

5.3.4 Cascade Effects Among Euro Area Sovereigns

When analyzing contagion in the financial system, we have to take into account its spatial dimension: a default event for a particular sovereign might affect perceptions about the default of another and their potential joint default might spread to others, in many cases – safer, sovereign borrowers. This may ultimately affect the default expectations regarding the entire sovereign system. In this section, we show

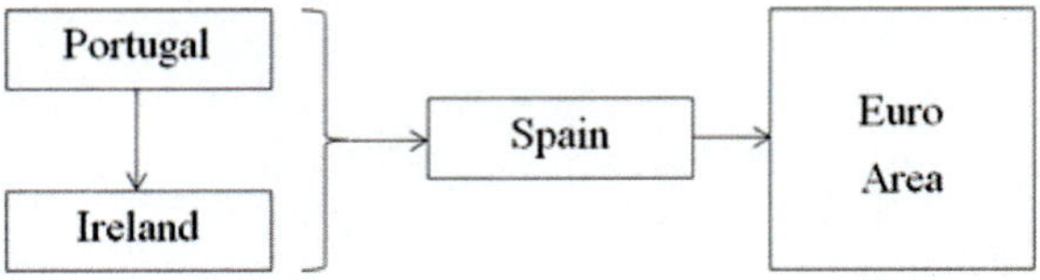

Fig. 5.5 Scheme for cascade effects in the euro area system of sovereigns. We assume that the default cascade starts from Portugal and in a first stage, we analyze how it affects Ireland. In a second stage, we examine how a joint default of these two countries could affect the solvency of Spain. In a third and final stage, we study the systemic risk contribution of a joint default of Portugal, Ireland and Spain on the default vulnerability of the EA system of sovereigns

how $\Delta CoJPoD$ can be used to investigate such "cascade" effects by considering a particular path, through which a default of one sovereign may spread within the system.

Figure 5.5 presents the hypothetical path we choose: We assume that the default cascade starts from Portugal and in a first stage, we investigate how it affects Ireland. In a second stage, we examine how a joint default of these two countries could affect the solvency of Spain. In a third and final stage, we study the systemic risk contribution of a joint default of Portugal, Ireland and Spain on the default vulnerability of the EA system of sovereigns. This cascade scenario was of particular interest during the Sovereign debt crisis in the euro area, since the bailouts of small countries like Greece, Ireland and Portugal affected the perceptions about the default risk of larger, troubled countries like Spain and Italy. In Table 4.5, we notice that the correlation between the CDS spreads of Portugal, Ireland and Spain is at a relatively high level, which makes a joint default very likely if any of them defaults. Alternatively, at the end of the section, we analyze the effect on systemic risk of a default of a large and safe country, such as Germany.

We start by examining the change in the conditional probability of default of Ireland if Portugal defaults. With a minor abuse of notation for the sake of parsimony, the general $\Delta CoJPoD$ definition in Eq. (5.3) transforms to:

$$\Delta CoJPoD_{Ir|Pt} = CoJPoD_{Ir|Pt} - PoD_{Ir}, \tag{5.9}$$

where $CoJPoD_{Ir|Pt}$ is the conditional probability of default of Ireland (Ir) given the assets of Portugal (Pt) cross their default threshold and PoD_{Ir} is the marginal (empirical) probability of default of Ireland. Compared to Eq. (5.3), the counterfactual joint probability of default of the surviving sovereigns in the system narrows down to a single dimension.

In Fig. 5.6, we present the results for each of the components on the right-hand side of Eq. (5.9). We witness a significant gap between the conditional probability of Ireland defaulting given Portugal defaults (blue) and its unconditional counterpart (red). Therefore, observing the individual probabilities of default, or even the raw individual CDS data, does not provide a comprehensive perspective on the complex interactions that underlie investors' perceptions about sovereign default risk. In

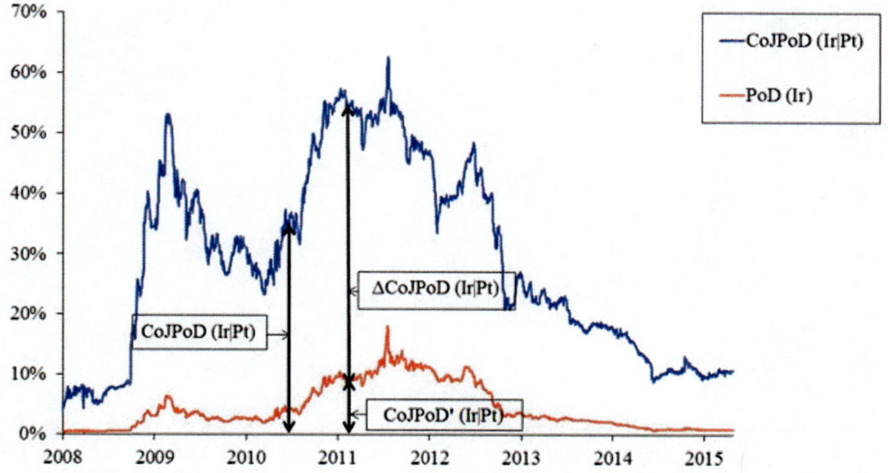

Fig. 5.6 Conditional probability of default ($\Delta CoJPoD$) of Ireland given Portugal defaults (blue) and unconditional probability of Ireland (red) for the period 01/01/2008–04/27/2015. Non-zero correlation structure is used in defining the prior distribution. The respective correlation coefficient is presented in Table 4.5. The correlation structure calculation is explained in the text and in Table 4.5. For an explanation of the abbreviations, see Fig. 4.4

Fig. 5.6, we observe that individual and conditional default risk decrease since their peak in 2011, and especially after Mario Draghi's "Whatever it takes" speech at the end of July 2012.

Figure 5.6 also exemplifies the contribution of our approach to modelling dependence under the CIMDO framework. If, as in Segoviano and Goodhart (2009), we had assumed zero correlation between the government assets of Portugal and Ireland, the conditional probability $CoJPoD_{Ir|Pt}$ would have equaled the unconditional probability $PoD_{Ir} = CoJPoD'_{Ir|Pt}$, yielding no additional information content stemming from the default of Portugal. Therefore, if we had relied only on conditional probabilities, observing the dynamics of $CoJPoD_{Ir|Pt}$ only would have delivered spurious impression of a rising spillover effect, when no information content is actually being captured. This point was shown in Sect. 5.3.3 and will be further elaborated upon in Sect. 5.5.1.

The blue line in Fig. 5.7 displays the dynamics of $\Delta CoJPoD_{Ir|Pt}$. We observe that the contribution of a Portuguese default to the distress vulnerability of Ireland rises from relatively modest 7 percentage points to almost 50 percentage points at the peak of the global recession after Lehman Brothers' bankruptcy. Although our measure declines afterwards, it hardly falls to 20 percentage points before 2010. We document a new rise from the beginning of 2010 onwards, reaching 50 percentage points in early 2011. The contribution slowly declines thereafter and stabilizes at around 20 percentage points by the end of 2012. Overall, we find a strong effect of a potential Portuguese default on the default expectations about Ireland.

Fig. 5.7 Change in the conditional probability of default ($\Delta CoJPoD$) of Ireland given Portugal defaults (blue) and the conditional probability of default ($\Delta CoJPoD$) of Spain, given Ireland and Portugal jointly default (red) for the period 01/01/2008–04/27/2015. Non-zero correlation structure is used in defining the prior distribution. The respective correlation coefficient is presented in Table 4.5. The correlation structure calculation is explained in the text and in Table 4.5. For an explanation of the abbreviations, see Fig. 4.4

The discussed results in Figs. 5.6 and 5.7 show that, as far as market perceptions are concerned, a default of Portugal is expected to substantially affect the default likelihood of Ireland. Yet, since both countries are relatively small, their difficulties could be fully addressed by the financial stabilization facilities, organized to prevent the spread of the Sovereign debt crisis: the European Financial Stability Facility (EFSF) and the European Financial Stabilization Mechanism (EFSM), both introduced in May 2010 with an initial mandate to borrow up to 500 billion Euros to maintain the financial stability of the eurozone. An additional amount of 250 billion Euros could be borrowed by the International Monetary Fund within this initial agreement. In December 2011, the size of public debt of Ireland and Portugal were 169 billion Euros and 184 billion Euros, respectively (Eurostat 2012) and hence their bailouts would have been fully covered by the aforementioned facilities. A recurring theme in the debate between policymakers and regulators at the time was whether a default of these smaller sovereigns could spread to the *larger* EA periphery economies, that is, to Italy and Spain. When we relate this to the general financial stability literature, Zhou (2010) points out that when assessing the systemic role of a financial institution, we should consider whether its distress co-occurs with distress of other institutions—the so called "too-many-to-fail" problem, investigated by Acharya and Yorulmazer (2007). Zhou (2010) argues that this effect is more relevant for financial crises than the popular "too-big-to-fail" argument. In the next step, we investigate how the joint default of Ireland and Portugal could affect the default perceptions about a default on Spanish government debt. To answer that question, we reformulate our Eq. (5.3) to take the following form:

$$\Delta CoJPoD_{Sp|Ir,Pt} = CoJPoD_{Sp|Ir,Pt} - PoD_{Sp}, \tag{5.10}$$

with $CoJPoD_{Sp|Ir,Pt}$ being the probability of Spain defaulting, given Ireland and Portugal default.

The results for $\Delta CoJPoD_{Sp|Ir,Pt}$ are represented by the red line in Fig. 5.7. Although the time series is less volatile, we observe similar level and time paths of default risk contribution to those of $\Delta CoJPoD_{Ir|Pt}$. The peak is in mid-2012 (50 percentage points), followed by a gradual reduction of the perceived default risk contribution up until July 2014 when the contribution rises again and plateaus at the relatively high level of 30 percentage points. Compared to a maximum of 8% individual probability of default of Spain in the observed time period, the contribution to the individual default risk of the country due to its interdependence with the two smaller countries ($CoJPoD_{Sp|Ir,Pt}$) is exceptionally high. Therefore, we find that a joint default of Ireland and Portugal substantially increases the perceived default risk of Spain. Although the levels subside after 2012, the figure reveals that Spain is still vulnerable to the risk of joint default of Portugal and Ireland. This could be attributed to the relatively better performance of the latter two countries in recent times, while the Spanish economy is still in a process of recovery from the debt crisis.

The final step of our cascade effects analysis is an examination of how a default of all three sovereigns affects the perceived default vulnerability of the EA system. This leads to the following $\Delta CoJPoD$ definition:

$$\Delta CoJPoD_{system-Sp,Ir,Pt|Sp,Ir,Pt} = CoJPoD_{system-Sp,Ir,Pt|Sp,Ir,Pt} - JPoD_{system-Sp,Ir,Pt}, \tag{5.11}$$

where $CoJPoD_{system-Sp,Ir,Pt|Sp,Ir,Pt}$ is the perceived probability of default of the surviving EA sovereigns, given a joint default of Spain, Ireland and Portugal.

In Fig. 5.8, we present the results for $\Delta CoJPoD_{system-Sp,Ir,Pt|Sp,Ir,Pt}$ in the sample period (red), compared to the respective $\Delta CoJPoD$ results when only Spain defaults in the EA system taken from Fig. 5.3 (black). We notice that the systemic default risk contribution triples during the post-Lehman global recession and doubles during the Sovereign debt crisis, compared to the case of a standalone default of Spain. Therefore, we argue that possible default cascade effects of reasonable size should be taken into account in policy decision-making in crisis times. Figure 5.8 confirms the decrease of the systemic risk contributions in the events of either Spain individually defaulting, or of Spain, Portugal and Ireland jointly defaulting. This is especially evident after the first quarter of 2013, which can mark the end of the Sovereign debt crisis for these countries.

Having investigated the propagation of default among the countries that were at the center of the Sovereign debt crisis either due to a shock to their financial systems (Ireland) or due to substantial structural economic problems (Portugal and Spain), an interesting question arises: What would be the effect of a default of a large and safe country? To this end, we substitute Portugal with Germany as a start of the default cascade. Figure 5.9 depicts the contribution of a default of Germany on the default risk of Ireland. The results are interesting and surprising: Despite the low probabil-

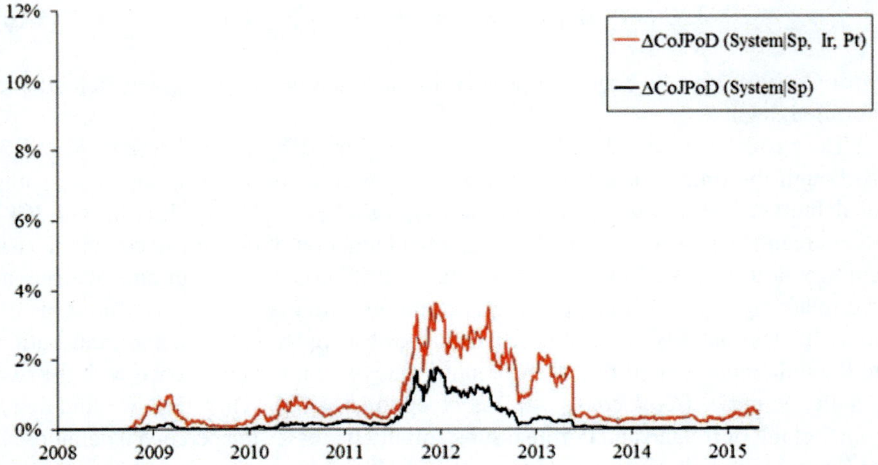

Fig. 5.8 Change in the conditional joint probabilities of default ($\Delta CoJPoD$) for the period 01/01/2008–04/27/2015. $\Delta CoJPoD(Sp, Ir, Pt)$ (red) is the conditional probability given Spain, Ireland and Portugal jointly default, while $\Delta CoJPoD(Sp)$ (black) is the respective conditional probability given a default of Spain, depicted in Fig. 5.3. Non-zero correlation structure is used in defining the prior distribution. The correlation matrix is presented in Table 4.5. The correlation structure calculation is explained in the text and in Table 4.5

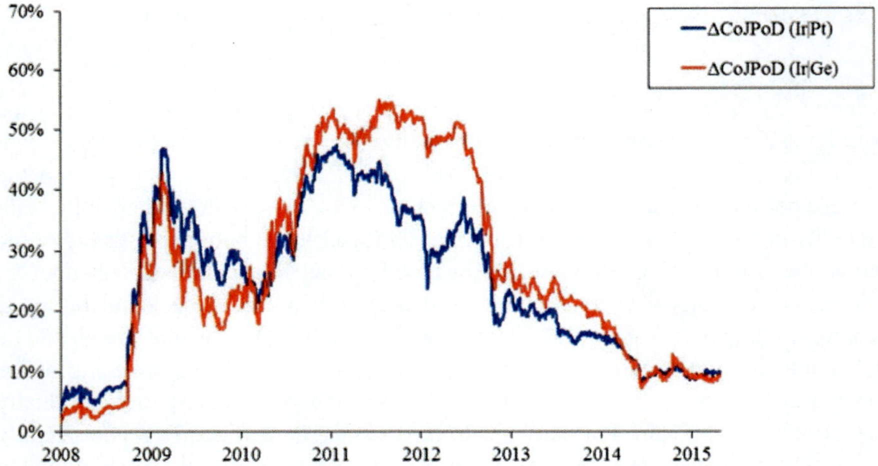

Fig. 5.9 Change in conditional probability of default ($\Delta CoJPoD$) of Ireland given Germany defaults (red) and change conditional probability of default ($\Delta CoJPoD$) of Ireland given portugal defaults (blue) for the period 01/01/2008–04/27/2015. Non-zero correlation structure is used in defining the prior distribution. The respective correlation coefficient is presented in Table 4.5. The correlation structure calculation is explained in the text and in Table 4.5

ity of default of Germany throughout the period, which should theoretically increase $\Delta CoJPoD$ (see Eq. (5.2)), in Fig. 5.9 we observe a similar effect on Ireland's default risk to the effect of a Portuguese default. This stems from the relatively low bilateral correlation between Germany and Ireland, compared to the correlation between Ireland and Portugal (see Table 4.5). The effect of the decrease in the denominator in Eq. (5.2) (due to the lower PoD of Germany) is counterbalanced by the decrease in the numerator due to the lower bilateral correlation.

In Fig. 5.10, we observe that, as expected, a joint default of Ireland and Germany is more likely to affect the default risk of Spain, compared to a joint default of Ireland and Portugal. The first reason for the larger contribution is the higher correlation between Germany and Spain. The second reason is the low individual default risk of Germany, compared to Portugal. Both effects are in the same direction and lead to an increase in $\Delta CoJPoD$. Figure 5.11 presents the contribution of a joint default of Germany, Ireland and Spain on the default risk of the euro area system. The effect on the systemic risk in the euro area is much larger than the effect of a joint default of Portugal, Ireland and Spain: Fig. 5.11 shows an up to eight times higher increase of $\Delta CoJPoD$ of the euro area. These results can again be traced back to Table 4.5: While the bilateral correlations between Germany and Ireland and Portugal (and Greece) are relatively low, Germany has the highest average bilateral correlation with the remaining countries in the euro area.

In summary, we find that defaults of small sovereigns may be highly correlated events and that their joint default can precipitate a default by a large sovereign and adversely affect the stability of the euro area. We also find that bilateral contagion

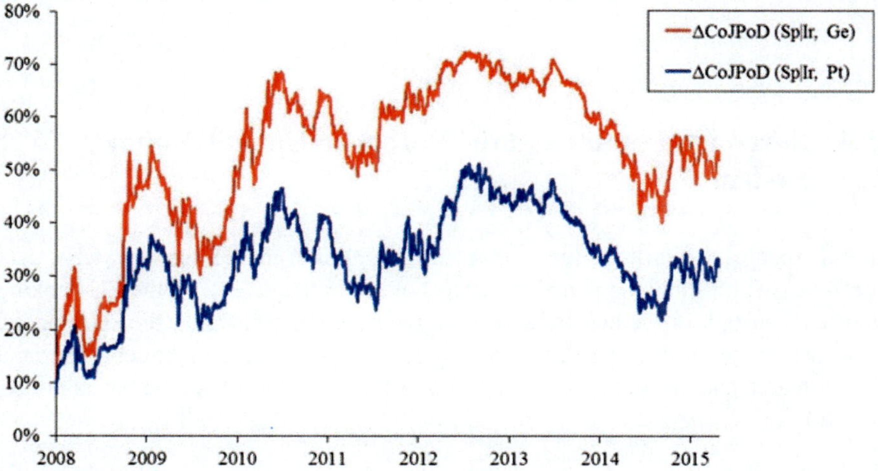

Fig. 5.10 Change in conditional probability of default ($\Delta CoJPoD$) of Spain, given Ireland and Germany jointly default (red) and change in conditional probability of default of Spain, given Ireland and Portugal jointly default (blue) for the period 01/01/2008–04/27/2015. Non-zero correlation structure is used in defining the prior distribution. The respective correlation coefficient is presented in Table 4.5. The correlation structure calculation is explained in the text and in Table 4.5

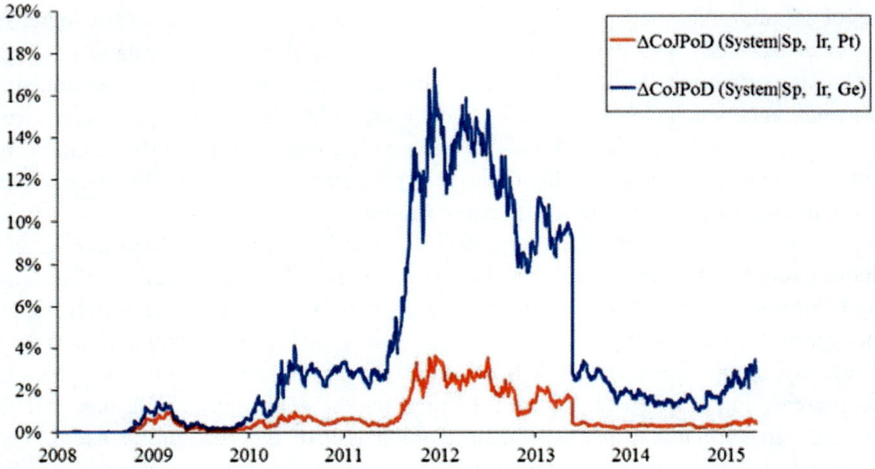

Fig. 5.11 Change in the conditional joint probabilities of default ($\Delta CoJPoD$) for the period 01/01/2008–04/27/2015. The blue line is the change in the conditional probability given Spain, Ireland and Germany jointly default, while the red line is the change in the conditional probability given Spain, Ireland and Portugal jointly default. Non-zero correlation structure is used in defining the prior distribution. The correlation matrix is presented in Table 4.5. The correlation structure calculation is explained in the text and in Table 4.5

effects of a default of a large, safe sovereign can be alleviated by low interconnectedness. However, the overall high correlation between large euro area countries would result in a much higher negative impact on the stability of the euro area system in that case.

5.4 Sovereign Default and the European Union Banking System

In this section, we shift the focus of our investigation from the pure sovereign debt perspective and study the perceived effects of sovereign default on the EU banking system. The topic of whether and how a sovereign default could affect the EU financial system is of major concern for regulators, as EU banks hold most of the debt generated by euro area countries and this debt accounts for a sizable part of the banks' assets portfolios.

5.4.1 Bank Data

In the analysis of spillover effects of sovereign default to the banking system, we use 36 European Union (EU) banks, out of which 28 are euro area banks. The lists of banks in our analysis are presented in Tables 5.4 and 5.5. The sources of the CDS data are CMA and Bloomberg.[14]

To reduce the effect of the curse of dimensionality, we choose to form equal-size portfolios within our banking sample. The latter construction choice makes our results comparable across portfolios. The allocation of each bank in a portfolio is governed by the position of the bank in the distribution of a set of financial characteristics. We select ten financial statement indicators, singled out in the financial literature as important systemic risk factors. Those factors form five broad groups: size, financial gearing, asset quality, performance, and liquidity and funding. The raw data for the individual bank characteristics for the analyzed period are provided by Bloomberg and Bankscope.

Size. We measure the bank's size by the amount of its total assets (TA). Adrian and Brunnermeier (2016) identify size as a major driver of systemic risk, according to the theory of the "margin spiral" (Brunnermeier and Pedersen 2009). The authors provide evidence that banks adjust their assets, such that leverage is high in upturns and low in downturns of the economic cycle, making leverage a procyclical characteristic. Sorting by size should provide us with insights whether bigger banks were exposed to higher default risk stemming from sovereign difficulties in the indicated period.

Leverage. Adrian and Brunnermeier (2016) propose the assets-to-equity (AE) ratio as a measure of financial gearing. Our hypothesis is that banks with higher leverage should be more susceptible to adverse credit events in the financial markets. Moreover, many large European Union banks invested heavily in EA sovereign bonds before and during the sovereign debt crisis (EBA 2011; IMF 2011) and could become insolvent in case of a sudden drop in the value of their assets.

Asset quality. We use two measures for asset quality. The first is the ratio of loan loss provisions to net interest income. This indicator reflects whether the lending risk undertaken by the banks is appropriately remunerated by higher interest margins. Hence, this measure should be as low as possible. Our second measure for the quality of a bank's portfolio of assets is the ratio of non-performing loans to total loans and is sometimes referred to in practice as the "doubtful loans" (DL) ratio. An increase in this measure should make banks more vulnerable to credit events that further impair their loan quality.

Performance. We use four indicators to measure a bank's performance. The first indicator is the return on equity (ROE), which is a standard measure of corporate efficiency. The main benefit of this measure is that it shows the profitability of the funds invested or reinvested in the company's equity. The main drawback comes from the fact that high-leverage companies could have artificially high ROE ratios, which

---

[14] CDS data for some of the big banks in the European Union was not available from CMA at the time of download. Therefore, we complement the CDS data for banks with data from Bloomberg to achieve a higher coverage of the EU banking system.

Table 5.4 List of euro area banks used in our analysis

Euro area banks		
	Country code	Name
1	AT	Erste Group Bank AG
2	AT	Raiffeisen Bank International Austria
3	BE	Dexia SA
4	BE	KBC Groep NV
5	DE	Bayerische Landesbank
6	DE	Commerzbank AG
7	DE	Deutsche Bank AG
8	DE	Landesbank Berlin Holding AG
9	ES	Banco Bilbao Vizcaya Argentaria
10	ES	Banco de Sabadell SA
11	ES	Banco Santander SA
12	FR	BNP Paribas
13	FR	Credit Agricole SA
14	FR	Natixis
15	FR	Societe Generale
16	IE	Allied Irish Banks PLC
17	IE	Governor & Co of the Bank of Ireland
18	IE	Irish Life and Permanent
19	IT	Banca Monte dei Paschi di Siena
20	IT	Banca Popolare di Milano
21	IT	Banco Popolare SC
22	IT	Intesa Sanpaolo SpA
23	IT	UniCredit SpA
24	NL	ING Groep NV
25	NL	Rabobank
26	NL	SNS Bank Netherlands
27	PT	Banco Comercial Portugues SA
28	PT	Espirito Santo Financial Group

might reflect the company's excessive risk-taking, rather than its growth potential. The second indicator is the return on assets (ROA), which is the profit from every euro of assets that the bank controls. A potential weakness of this accounting measure is that the balance sheet value of assets may differ from the market value of assets, making it difficult to draw comparisons across industries. Within the banking industry this is less of an issue, due to the relatively regular marking to market of assets. Our third measure is the net interest margin (NIM). The NIM is calculated as interest income minus interest expenses over average earning assets. It indicates how successful the bank's investment decisions were in comparison to the interest-bearing

Table 5.5 List of additional non-euro area European Union banks used in our analysis

Other European Union banks

	Country code	Name
1	DK	Danske Bank A/S
2	GB	Barclays PLC
3	GB	HSBC Holdings PLC
4	GB	Lloyds Banking Group PLC
5	GB	Royal Bank of Scotland Group
6	GB	Standard Chartered PLC
7	SE	Nordea Bank AB
8	SE	Skandinaviska Enskilda Banken

assets. A negative value could indicate a non-optimal banking credit policy or a fast deterioration in the quality of assets. Our last measure is the bank's efficiency ratio (ER). This ratio compares the overhead costs of running the bank to the revenues from the bank's business. The higher the ratio, the less efficient the bank's operations are.

Liquidity and Funding. Our last category includes two indicators. The first is the deposits-to-funding (DF) ratio, which is calculated by dividing total deposits by total funding (sum of total deposits, short- and long-term borrowing and repurchase agreements). This measure reflects the share of stable funding (deposits) to the total amount of a bank's funding. The less a bank relies on wholesale funding, the less exposed it is to global volatility and credit crunches during global crises. The higher this ratio, the better protected a bank is against global market fluctuations.[15] The second liquidity measure is net loans to total assets. This liquidity indicator reflects the share of loans less loan loss provisions to total assets. An increase in this ratio may signal liquidity shortages, as the company has an increasing proportion of illiquid assets (loans).

The frequency of the financial characteristics is quarterly for Bloomberg and annual for Bankscope data. In Table 5.6, we present the ranking of banks according to the ten factors.

5.4.2 Estimation Strategy

We choose a particular sovereign, Spain, to be the trigger of default risk in the banking system. As previously argued, due to their small relative size, it is safe to assume that Greece, Ireland and Portugal could be bailed out if needed and hence the resulting

[15] Of course, this measure is only meaningful when there are no bank runs. Since bank runs will affect not only the deposits, but also the general funding availability, the information content of this liquidity measure is reduced during such periods.

Table 5.6 Ranking assignment in descending order of the 36 banks used in our analysis with respect to 10 financial characteristics. The numbers in the columns for the financial characteristics come from the ordering in Tables 4.1 and 5.5. The values from 29 to 36 are given to the non-euro area European Union banks in the same order as in Table 5.5. PF 1 to PF 9 list the 4 banks that are included in the final 5-entity portfolios for each financial characteristic. The abbreviations stand for total assets (TA), return on equity (ROE), return on assets (ROA), net interest margin (NIM), efficiency ratio (ER), deposits-to-funding (DF) ratio, assets-to-equity (AE) ratio, loan-loss-provisions-to-net-interest-income (LLP-to-NII) ratio, non-performing-loans-to-total-loans ("doubtful loans", DL) ratio, net-loans-to-total-assets (NL-to-TA) ratio

		Financial characteristics									
Ranking	Portfolios	TA	ROE	ROA	NIM	ER	DF	AE	LLP-to-NII	DL	NL-to-TA
1	PF 1	33	11	2	2	4	31	3	16	16	26
2		12	23	8	1	16	34	8	17	21	27
3		7	33	11	9	14	18	7	18	19	10
4		31	31	10	11	29	24	18	32	23	25
5	PF 2	30	7	9	34	8	1	26	3	17	20
6		13	9	35	20	6	7	24	26	22	17
7		24	34	22	19	20	23	6	33	1	21
8		11	30	19	31	32	19	13	5	32	19
9	PF 3	15	32	1	22	21	4	29	27	2	16
10		23	22	36	10	19	2	30	31	8	28
11		32	12	30	23	33	25	5	14	6	2
12		6	25	20	21	13	32	12	30	15	9
13	PF 4	22	15	31	16	1	27	32	24	33	32
14		25	6	34	4	22	11	17	21	30	1
15		3	4	27	27	24	9	14	6	12	11
16		9	1	12	28	18	20	36	15	13	23
17	PF 5	35	35	24	32	15	10	35	1	10	22
18		14	29	15	17	25	16	15	23	4	29
19		29	19	28	25	27	17	16	2	26	3
20		5	2	23	30	26	33	33	10	9	35
21	PF 6	4	3	6	33	23	36	4	12	14	18
22		34	13	21	35	28	21	31	11	29	24
23		36	24	7	29	12	26	9	9	27	36
24		19	21	25	26	36	35	25	28	11	5
25	PF 7	1	36	29	12	30	6	10	19	31	34
26		17	27	13	24	2	22	11	13	5	6
27		16	10	32	13	34	30	34	22	7	4
28		8	28	14	15	31	13	23	20	20	31
29	PF 8	21	20	18	36	35	28	27	29	18	33
30		2	8	5	6	10	29	1	36	34	15
31		27	14	26	7	5	15	19	4	25	8
32		10	18	17	3	9	8	28	35	24	12
33	PF 9	26	26	4	8	11	5	20	25	28	13
34		28	17	3	5	17	12	2	34	36	30
35		18	5	33	14	7	3	22	8	35	14
36		20	16	16	18	3	14	21	7	3	7

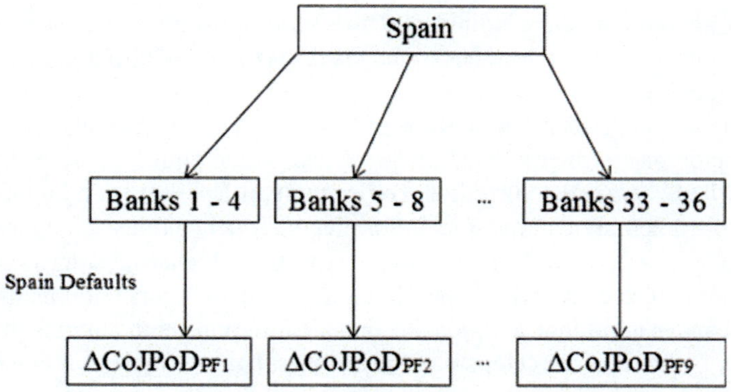

Fig. 5.12 Scheme for construction of portfolios to examine spillover effects of sovereign default to the banking system. We start by sorting our set of 36 banks by the time average of each of the financial characteristics in Sect. 5.4. The banks are listed in Tables 4.1 and 5.5 and the results from the sorting are presented in Table 5.6. To address the "curse of dimensionality," we divide the 36 banks into 9 subsets, resulting in 4 banks per (sub)portfolio. In order to examine the effect of a sovereign default on the banking system in this framework, we need to explicitly include a sovereign (Spain) in each subportfolio, which will act as a trigger for sovereign default in the respective portfolio. Using this strategy, we reduce the joint density modeling to a 5-dimensional problem. For each portfolio within each characteristic, we consider the $\Delta CoJPoD$ in case of Spain's default

default risk within the EU banking system could be relatively easily defused. That leaves Spain and Italy as the main concern among the GIIPS (Greece, Ireland, Italy, Portugal and Spain). The debt level of these two countries might make it infeasible to prevent a default event if they meet difficulties to service their payments (e.g. due to short-term illiquidity issues). For this reason, the ECB has continuously intervened in the debt market once Spain and Italy announced that they would issue new debt to cover their short-term funding needs. To relate to the example from the section on cascade effects, in this section we analyze how a default of Spain could affect the market perceptions about risk in the banking system.

We start by sorting our set of banks by the time average of each of the financial characteristics in the previous section. To address the "curse of dimensionality," we then divide the 36 banks into 9 subsets, resulting in four banks per (sub)portfolio.[16] According to our definition of the financial system, the system could be represented by a portfolio of a set of entities, and it is independent from the remaining entities, which are not included in the system. Therefore, in order to examine the effect of a sovereign default on the banking system in this framework, we need to explicitly include a sovereign (Spain) in each subportfolio, which will act as a trigger for sovereign default in the respective portfolio. Figure 5.12 demonstrates the construction of each portfolio. Using this strategy, we reduce the joint density modeling to a 5-dimensional problem. For each portfolio within each characteristic, we consider the $\Delta CoJPoD$

[16] We "fix" the composition of the subportfolios in favour of a dynamic composition, in order to make the multivariate probability results comparable across periods.

in case of Spain's default, resulting in 90 time series for further analysis. For the ease of exposition, we present averages of the results for three subgroups: portfolios 1 to 3, 4 to 6 and 7 to 9.

The main hypothesis in the analysis in this section is that if a financial characteristic is important for international investors, we should notice a particular ordering of $\Delta CoJPoD$ across the portfolios. For example, if leverage is an important characteristic for international investors in forming their perceptions about banking susceptibility to sovereign default, we would expect that higher-leveraged banks react more strongly to such an event. Therefore, in solving the "curse of dimensionality," we also manage to provide additional economic intuition behind our results. This is, of course, by no means a ceteris paribus analysis, but nonetheless, it could provide useful policy implications and important insights for further research.

5.4.3 Spillover Effects from Sovereigns to Banks

Figure 5.13 depicts the $\Delta CoJPoD$ results given a default of Spain for 9 portfolios sorted by *size*.[17] The results show a clear ordering—we notice a split of our portfolios in two groups, with the biggest banks in our sample reacting much more strongly to increases in Spanish default risk. Several spikes occur throughout 2008 up to the end of the global recession in mid-2009. After relatively stable 9 months, the conditional sovereign default contribution to the fragility of the biggest banks rises again in March–April 2010; and in mid-2011 it surpasses the levels during Lehman Brothers' turmoil. The higher level after July 2011 could be attributed to increased attention of markets to the problems of Italy and Spain. Our results could be explained not only by the sizable EA sovereign debt holdings on the balance sheets of the biggest banks, but also by the uncertainty about the economic conditions in the European Union during the sample period. The high susceptibility of big banks to sovereign default risk might be related to "too-big-to-save" (Hellwig 1998) considerations by international investors. The recent experience with the prolonged political process of bailout-package ratifications might explain why investors could be skeptical about multilateral government cooperation to support these international conglomerates.

With regard to *leverage*, Fig. 5.14 provides a rather mixed picture. There are significant peaks during the sample period, especially in the second half of 2011, but the most vulnerable bank groups turn out to be those with relatively modest levels of leverage. This indicates that financial gearing might not be a good indicator for the reaction of banks to sovereign debt problems. An argument why leverage can provide misleading results is the fact that during crises financial institutions tend to procyclically reduce their leverage level, sometimes at a high cost, which makes them highly vulnerable to financial market volatility. A more reasonable explanation arises when we observe the actual composition of portfolios 7 to 9 in column AE

[17] We present and interpret the results for several financial characteristics only. The rest of the results are available upon request.

Fig. 5.13 Change in the conditional joint probabilities of default ($\Delta CoJPoD$) involving 9 portfolios of 36 banks and 1 sovereign for the period 01/01/2008–12/31/2011. Each portfolio includes 4 banks and 1 sovereign. Spain is considered uniformly as a default-risk-triggering sovereign. Only the series conditioning on default of Spain are presented for each portfolio. The assignment of a bank to a portfolio is governed by its **size**, compared to the rest, sorted in descending order. E.g. Portfolio 1 (PF 1) includes the four banks with the highest value of *total assets*. For presentation purposes, the **average results of the top, middle and bottom 3 portfolios** are shown in the figure. The set of banks is presented in Tables 4.1 and 5.5. The complete portfolio assignment is presented in Table 5.6, column TA. Non-zero correlation structure is used in defining the prior distribution. The correlations are calculated between changes in the 5-year CDS spreads of the respective banks and sovereigns in each portfolio. $\Delta CoJPoD$ is calculated analogously to the sovereign case. The total assets are denominated in euro

of Table 5.6. They consist primarily of Spanish, Italian and Portuguese banks that have high exposure to the debt of their own governments, who incidentally were at the center of the sovereign debt crisis. As Sect. 5.3.4 showed, the destiny of these governments' debt is very highly correlated and the effect of a prospective Spanish default on Spanish banks would be indicative for the effect of Italian or Portuguese government defaults on their respective banks. The remaining two groups of portfolios (1 to 3 and 4 to 6) have the more intuitive ordering that more levered banks are more vulnerable to government default (see, e.g., Acharya and Thakor 2016).

Interestingly enough, the sorting by return on equity (Fig. 5.15) reveals that the market perceptions of the default risk of the *highest-performing* banks tend to react more intensively to sovereign default risk. The top three portfolios appear to have four to six times higher $\Delta CoJPoD$ than the second subgroup, especially in the periods around the Bear Stearns episode, the bankruptcy of Lehman Brothers and the following global recession, as well as during the more recent events, related to the sovereign debt crisis. A possible explanation is that in international investors' view

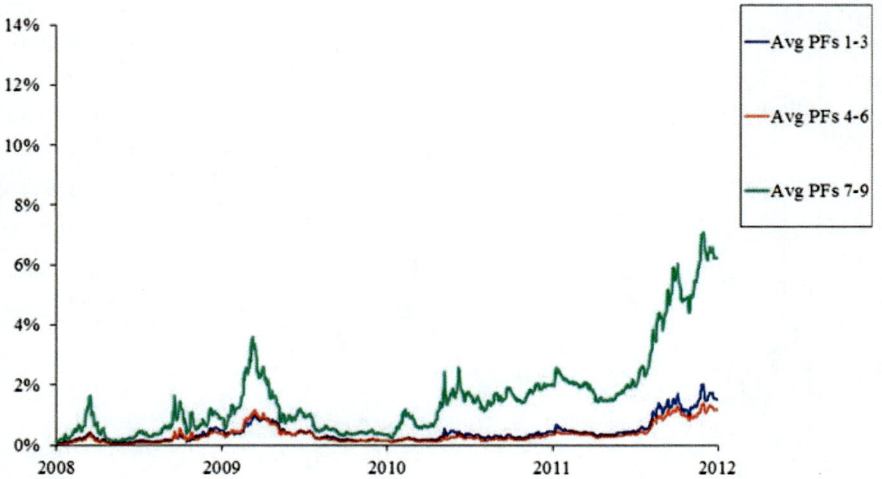

Fig. 5.14 Change in the conditional joint probabilities of default ($\Delta CoJPoD$) involving 9 portfolios of 36 banks and 1 sovereign for the period 01/01/2008–12/31/2011. Each portfolio includes 4 banks and 1 sovereign. Spain is considered uniformly as a default-risk-triggering sovereign. Only the series conditioning on default of Spain are presented for each portfolio. The assignment of a bank to a portfolio is governed by its **leverage**, compared to the rest, sorted in descending order. E.g. Portfolio 1 (PF 1) includes the four banks with the highest value of *assets to equity*. For further details, see the caption of Fig. 5.13 and the text

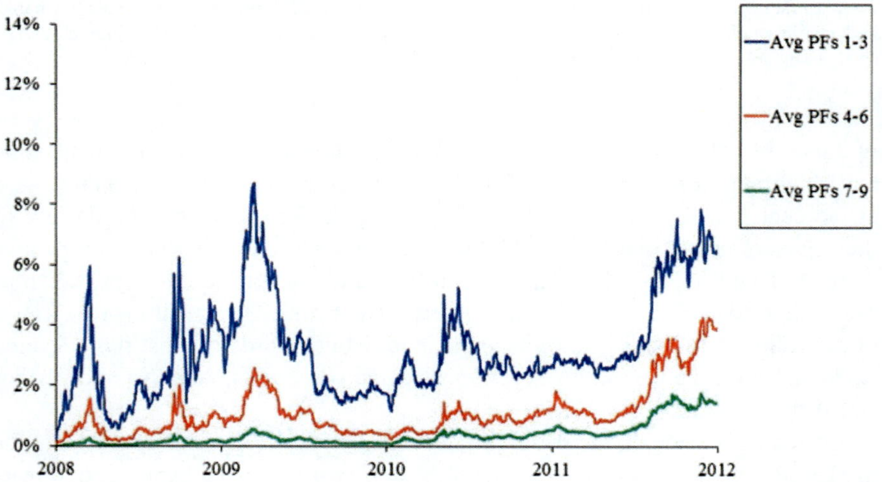

Fig. 5.15 Change in the conditional joint probabilities of default ($\Delta CoJPoD$) involving 9 portfolios of 36 banks and 1 sovereign for the period 01/01/2008–12/31/2011. Each portfolio includes 4 banks and 1 sovereign. Spain is considered uniformly as a default-risk-triggering sovereign. Only the series conditioning on default of Spain are presented for each portfolio. The assignment of a bank to a portfolio is governed by its **performance**, compared to the rest, sorted in descending order. E.g. Portfolio 1 (PF 1) includes the four banks with the highest value of *return on equity*. For further details, see the caption of Fig. 5.13 and the text

Fig. 5.16 Change in the conditional joint probabilities of default ($\Delta CoJPoD$) involving 9 portfolios of 36 banks and 1 sovereign for the period 01/01/2008–12/31/2011. Each portfolio includes 4 banks and 1 sovereign. Spain is considered uniformly as a default-risk-triggering sovereign. Only the series conditioning on default of Spain are presented for each portfolio. The assignment of a bank to a portfolio is governed by its **asset quality**, compared to the rest, sorted in descending order. E.g. Portfolio 1 (PF 1) includes the four banks with the highest value of *doubtful loans*. For further details, see the caption of Fig. 5.13 and the text

the higher performance signals that the banks in question are involved in activities that are too risky (see, e.g. Delis and Kouretas 2011).

We now turn our attention to the sorting by *asset quality*, measured by the doubtful loans ratio. Figure 5.16 provides some evidence that international investors do take asset quality into account when assessing default risk. The middle set of portfolios is consistently above the top and bottom set, but since April 2010 the banks with the highest doubtful loans ratio gradually reduce the differential. Therefore, it could be the case that in the second half of our sample period this factor gains increasing importance for international investors.

The ordering by total deposits to total funding (Fig. 5.17) follows our expectations that banks with lower values for this indicator (hence more reliant on funding from financial markets) are more vulnerable to sovereign default risk. The "correct" ordering of the portfolios is especially evident after the outbreak of the sovereign debt crisis.

Our analysis leads us to conclude that bank size, leverage, riskiness of operations, asset quality and liquidity and funding are important factors that affect the transmission of default risk from EA sovereigns to EU banks are.

Fig. 5.17 Change in the conditional joint probabilities of default ($\Delta CoJPoD$) involving 9 portfolios of 36 banks and 1 sovereign for the period 01/01/2008–12/31/2011. Each portfolio includes 4 banks and 1 sovereign. Spain is considered uniformly as a default-risk-triggering sovereign. Only the series conditioning on default of Spain are presented for each portfolio. The assignment of a bank to a portfolio is governed by its **liquidity and funding**, compared to the rest, sorted in descending order. E.g. Portfolio 1 (PF 1) includes the four banks with the highest value of *total deposits to total funding*. For further details, see the caption of Fig. 5.13 and the text

5.5 Comparison with Common Probability Measures

5.5.1 Comparison with the Probability Measures by Segoviano and Goodhart (2009)

As mentioned in Sect. 5.2, the existing probability measures of systemic risk may offer misleading information about contagion and spillover effects, since the authors focus primarily on conditional probabilities and not on contributions to default risk that assess the information content in the default of a sovereign or a bank. In this subsection, we provide empirical comparison of the $\Delta CoJPoD$ to the probability of at least one bank to default, given a certain bank defaults, or PAO, introduced by Segoviano and Goodhart (2009). Radev (2014) shows that a generalized version of this measure, the probability of all banks to default, given a bank defaults is exactly the CoJPoD in Eq. (5.2) and hence that version of the PAO is directly comparable to the measure that we introduce. For the purpose of the comparison, we translate the measure to sovereigns. As mentioned above, Segoviano and Goodhart (2009) assume zero correlation between the entities under investigation and interpret the dynamics of the resulting series as a probability of spillover effects. In Sect. 3.2.2, we showed that assuming zero correlation in a CIMDO setting effectively means that we assume the underlying entities to be independent. Also, as argued in Sect. 5.2, in this case the default contribution of any of the sovereigns is not different from 0, due to the fact that under independence, the conditional and unconditional JPoDs are the same. These theoretical considerations were illustrated in Fig. 5.4, where $\Delta CoJPoD$ for all countries is practically zero. Therefore, the analysis of $CoJPoD$ alone, as would be suggested by Segoviano and Goodhart (2009), cannot provide any additional information stemming from the default of a particular sovereign, since no contagion effects can be captured.

5.5.2 Comparison with Other Measures

Bisias et al. (2012) provide an extensive overview and classification of most of the existing (at the time) systemic risk measures. By definition, the $\Delta CoJPoD$ is related to two main types of measures of systemic risk: on the one hand, to general systemic risk contributions measures, like the CoVaR and the Shapley value, and on the other, to measures of joint default risk of financial institutions, like the unconditional probability of at least two banks to default simultaneously (Avesani et al. 2006; Radev 2014) and the conditional probability of at least one bank to default if another bank defaults (Segoviano and Goodhart 2009; Radev 2014). This section compares the $\Delta CoJPoD$ to these related measures.

Since the $\Delta CoJPoD$ is a measure of default *risk contributions*, the first candidate for a comparison is the $\Delta CoVaR$ (Adrian and Brunnermeier 2016), which compares the value-at-risk of the financial system if an institution's return is in the extreme negative tail of its distribution (crisis period) to the value-at-risk of the system if the aforementioned institution is at the median of its return distribution.[18] Gramlich and Oet (2011) outline several properties that a successful systemic risk measure should possess: consistency, flexibility, a forward-looking focus, correspondence with empirical data, suitability for the need of financial regulators. In that respect, as mentioned earlier, the CoVaR suffers from the limited number of extreme returns in market data, which makes it less suitable for a consistent forward-looking estimation of systemic risk. Since we base our measure on CDS data, it is exclusively forward-looking and is consistent with the default expectations of market participants within a 5-year horizon. Another limitation of the CoVaR is that it is unable to tackle multivariate interactions, as it models either interactions between two institutions, or between an institution and an aggregated index of the financial system. Therefore, it neglects the underlying dependence between the institutions in the system's "portfolio". In contrast, in our approach we employ a "true" multivariate setting for our $\Delta CoJPoD$ measure by explicitly modeling the dependence structure among all sovereigns in the system. Furthermore, our empirical applications showed how

[18] We should note that our definition of the counterfactual state of the world differs from the median state used for the final version of the $\Delta CoVaR$. In fact, or measure is related to an earlier version of the latter, where an institution in distress is excluded from the counterfactual. In our probability setting, this earlier concept is more appropriate in identifying the *information content* from a default.

flexible the measure is in capturing not only individual default contributions, but also cascade effects from a simultaneous default of several countries.

The Shapley value (SV) may be considered as an umbrella term for all systemic risk measures, since it can use any systemic risk contribution measure as a basis. It calculates an average value of all possible permutations of the difference in risk between coalitions within the financial system that include a particular institution and coalitions that exclude it. The notion of the SV comes from the game theory literature, where it is used to measure the fair allocation of gains within a coalition. The SV has a number of desired properties: the total gain of the coalition is distributed (Pareto efficiency); players with the same contribution to the gains have the same SV (symmetry); the individual contributions add up to the total gain (additivity); a player with no marginal contribution has a SV of zero (zero player). The $\Delta CoJPoD$ shares many similarities to the SV, since the choice of the "correct" sub-system is at the basis of the derivation of any of the measures in the extended $\Delta CoJPoD$ family. The difference here is that the focus is on a particular permutation of the set of players/countries and not on an averaged value of all possible permutations. The estimation procedure of $\Delta CoJPoD$ can easily be extended to meet the general requirements for a Shapley value measure.

With respect to the estimation of multivariate PoDs, financial stability literature puts an emphasis mainly on measuring distress risk of banks (see, for example, Lehar 2005; Avesani et al. 2006; Segoviano and Goodhart 2009; Giglio 2016). Of particular interest to supervisors is the probability of at least two banks defaulting (see Avesani et al. 2006; Radev 2014 for two approaches to derive this measure). Gorea and Radev (2014) and Radev (2014) propose procedures to calculate these measures in a sovereign context. The latter application is rarely explored in the literature, due to the different nature of sovereign and corporate assets. As in the case of the current work, the latter two papers do not try to define sovereign assets (contrary to the approach of, for instance, Gray 2011), but rather rely on the judgement of market participants about the probability of a latent sovereign assets process to cross a given sovereign default threshold. The latter probability is derived from CDS spreads. This agnostic view allows researchers to extend the definition of sovereign assets to anything that would affect a government's decision to default, be it public revenues or willingness to pay.

Most of the proposed probability measures in the cited studies are unconditional by definition and when authors derive conditional measures, they do not compare them to counterfactual events. As we show in Sect. 5.3, this can lead to spurious estimation of spillover effects, since the actual additional information provided by these measures is not clear. A notable exception is the "spillover component" or "contagion effect" measure, derived by Lucas et al. (2014). In contrast to the approach underlying $\Delta CoJPoD$, however, the authors choose to take the difference between the conditional probability of default of a sovereign (Portugal in their case), given another sovereign (Greece) defaults, and the conditional probability given the latter does not default. The latter term is the authors' definition of the "tranquil" state of the world. The authors keep the same dependence structure in both components of their measure, which prevents them from interpreting it as a contribution to systemic default risk due to interdependence. Moreover, the "spillover component" is a bivariate measure, investigating the effect of a default of a sovereign on another sovereign and the authors do not explore its implications for systemic risk. For the sake of comparing it to $\Delta CoJPoD$, we generalize it to the systemic context. The "spillover component" can then be represented in the following way:

$$\Delta SpilloverComponent_{system_{-k}|x_k>\bar{\mathbf{x}}_k} = CoJPoD_{system_{-k}|x_k>\bar{\mathbf{x}}_k} - CoJPoD''_{system_{-k}|x_k<\bar{\mathbf{x}}_k}, \tag{5.12}$$

where $CoJPoD''_{system_{-k}|x_k<\bar{\mathbf{x}}_k}$ is the conditional joint probability of default of the euro area if sovereign k does not default. The "spillover component" can then be translated to the observed difference in average outcomes on the left-hand side of Eq. (5.7), as opposed to the ATET that $\Delta CoJPoD$ measures. Such definition may lead regulators to interpret the "spillover component" as a measure of the risk of the system if a sovereign is allowed to default, relative to the case when it is *bailed out*. This is at odds with the general modeling of bailout probability in the financial stability literature, where higher bailout expectations affect the expected default risk in a downward direction. Hence, higher bailout expectations should reduce CDS spreads. Since CDS spreads are used for the calculation of both the conditional probability and its counterfactual, the effect of bailout expectations remains unclear, even after estimating the "spillover component". As outlined in the previous section, the $\Delta CoJPoD$ has an intuitive interpretation and we suggest applying this measure for regulatory purposes, after explicitly adjusting CDS spreads for bailout expectations in the first step of the procedure.[19] This extension of the procedure remains beyond the scope of the current book.

5.6 Conclusion

This chapter introduces a new way of measuring systemic risk contributions—not in terms of losses to the system as in the traditional financial stability literature, but rather in terms of conditional probabilities of default. Thus, we bridge the gap between two strands of literature: the literature on systemic risk contributions (Adrian and Brunnermeier 2016; Acharya et al. 2009, 2012; Tarashev et al. 2010; Brownlees and Engle 2017) and the literature on multivariate default risk measurement (Gray et al. 2007; Lucas et al. 2014; Gorea and Radev 2014; Radev 2014).

The new systemic risk measure that we propose, the $\Delta CoJPoD$, assesses the information content of a sovereign default and the effect of interdependence on the

[19] Furthermore, since regulators can either observe a default or a non-default, and never both counterfactuals, the calculation of the ATE suffers from a selection bias (see, Angrist and Pischke 2009, p. 22), since a sovereign can self-select to default or not and is therefore of limited interest for causal analysis. This, however, is more of a statistical issue, rather than probabilistic, and both types of systemic risk contributions can be readily calculated, once we recover the overall sovereign asset distribution.

general default risk of the financial system. The measure is related to the CoVaR, the Shapley value, the Marginal Expected Shortfall and SRISK and captures the relationship between the overall systemic fragility and the individual default risk. We apply our measure to two cases: first, we estimate individual default contributions of euro area sovereigns to the systemic default risk in the euro area. Second, we investigate cascade effects among euro area sovereigns.

Our results suggest that interdependence plays a major role in investors' perceptions about systemic risk in the European financial system. We find that countries with a relatively small size, such as the Netherlands, might have a significant systemic risk contribution if investors perceive them to be interdependent with others in the euro area. Another important conclusion from our analysis is that once a sovereign default is triggered, it is very likely to be followed by more sovereign defaults if sovereign assets are highly correlated. The joint default of two relatively small sovereigns, such as Ireland and Portugal, increases the probability of default of Spain by up to 50 percentage points. This effect has been persistent since Lehman's collapse. Therefore, our results provide support for the determined and often costly efforts of the European regulators and policy-makers to prevent a sovereign default. Bilateral contagion effects can be alleviated by low interconnectedness between countries, but the aggregate impact on the stability of the euro area system of sovereigns depends on the overall interconnectedness of sovereigns within the system.

We should note that $\Delta CoJPoD$ could be applied in any context where PoDs of sovereigns, companies or industries can be estimated. This includes a much wider range of scenarios than the investigated in this chapter. For instance, the measure can be used to investigate the spillover effects from sovereigns to financial institutions, or among industries within a country or a region. Convinced by our approach, the ECB started to use unconstrained correlation estimates for their bank risk measures based on minimum cross-entropy, improving on the zero-correlation assumptions they had made before. Therefore, we believe that $\Delta CoJPoD$ is a useful ingredient in the decision-making process of regulators and policy makers.

References

Acharya, V., Thakor, A.: The dark side of liquidity creation: leverage and systemic risk. J. Financ. Intermed. **2016**, 4–21 (2016)

Acharya, V., Yorulmazer, T.: Too many to fail–an analysis of time-inconsistency in bank closure policies. J. Financ. Intermed. **16**, 1–31 (2007)

Acharya, V., Pedersen, L., Philippon, T., Richardson, M.: Regulating systemic risk. In: Acharya, V., Richardson, M. (eds.) Restoring Financial Stability: how to Repair a Failed System. Wiley, New York (2009)

Acharya, V., Pedersen, L., Philippon, T., Richardson, M.: Measuring systemic risk. CEPR Discussion Papers 8824. C.E.P.R. Discussion Papers (2012)

Adrian, T., Brunnermeier, M.: CoVaR. Am. Econ. Rev. **106**(7), 1705–1741 (2016)

Angrist, J., Pischke, J.-S.: Mostly Harmless Econometrics: an Empiricist's Companion. Princeton University Press, Princeton (2009)

Avesani, R., Li, J., Pascual, A.: A new risk indicator and stress testing tool: a multifactor Nth-to-default CDS basket. IMF Working Papers 06/105. International Monetary Fund (2006)

Bisias, D., Flood, M., Lo, A., Valavanis, S.: A survey of systemic risk analytics. Working Paper 0001. U.S. Department of Treasury, Office of Financial Research (2012)

Brownlees, C., Engle, R.F.: SRISK: a conditional capital shortfall measure of systemic risk. Rev. Financ. Stud. **30**(1), 48–79 (2017)

Brunnermeier, M., Pedersen, L.: Market liquidity and funding liquidity. Rev. Financ. Stud. **22**(6), 2201–2238 (2009)

Delis, M., Kouretas, G.: Interest rates and bank risk-taking. J. Bank. Financ. **35**(4), 840–855 (2011)

EBA: 2011 EU-wide stress test aggregate report. European Banking Authority (2011)

Eurostat: Provision of deficit and debt data for 2011 - first notification. Newsrelease 62/2012. Eurostat (2012)

Giglio, S.: Credit default swap spreads and systemic financial risk. ESRB Working Paper Series 15. European Systemic Risk Board (2016)

Gorea, D., Radev, D.: The euro area sovereign debt crisis: can contagion spread from the periphery to the core? Int. Rev. Econ. Financ. **30**, 78–100 (2014)

Gramlich, D., Oet, M.: The structural fragility of financial systems: analysis and modeling implications for early warning systems. J. Risk Financ. **12**(4), 270–290 (2011)

Gray, D.: Enhanced bank stress testing incorporating sovereign risk, funding cost, and liquidity risk using contingent claims analysis. Background Paper. ESRB Advisory Technical Committee Workshop (2011)

Gray, D., Bodie, Z., Merton, R.: Contingent claims approach to measuring and managing sovereign risk. J. Invest. Manag. **5**(4), 5–28 (2007)

Hellwig, M.: Too big to be rescued? Schweiz. Bank **11**, 1–35 (1998)

Hull, J., White, A.: Valuing credit default swaps I: no counterparty default risk. J. Deriv. **8**(1), 29–40 (2000)

IMF: Global financial stability report. International Monetary Fund, Washington, DC (2011)

Lehar, A.: Measuring systemic risk: a risk management approach. J. Bank. Financ. **29**(10), 2577–2603 (2005)

Lucas, A., Schwaab, B., Zhang, X.: Conditional probabilities and contagion measures for euro area sovereign default risk. J. Bus. Econ. Stat. **32**, 271–284 (2014)

Lucas, A., Schwaab, B., Zhang, X.: Modeling financial sector joint tail risk in the euro area. J. Appl. Econom. **32**, 171–191 (2017)

Merton, R.: On the pricing of corporate debt: the risk structure of interest rates. J. Financ. **29**(2), 449–470 (1974)

Peña, J., Rodriguez-Moreno, M.: Systemic risk measures: the simpler the better. J. Bank. Financ. **37**(6), 1817–1831 (2013)

Radev, D.: Assessing systemic fragility: a probabilistic perspective. SAFE Working Paper 70. Research Center SAFE, Goethe University Frankfurt (2014)

Segoviano, M.: Consistent information multivariate density optimizing methodology. FMG Discussion Papers 557. International Monetary Fund (2006)

Segoviano, M., Goodhart, C.: Banking stability measures. IMF Working Paper 09/4. International Monetary Fund (2009)

Tarashev, N., Borio, C., Tsatsaronis, K.: Attributing systemic risk to individual institutions. BIS Working Papers 308. Bank for International Settlements (2010)

Zhou, C.: Are banks too big to fail? Measuring systemic importance of financial institutions. Int. J. Cent. Bank. **6**(34), 205–250 (2010)

Chapter 6
Summary and Final Words

The risk of default of sovereigns and banking institutions has proved to be of major concern for regulators during the Global financial and the recent euro area sovereign debt crises. In such turbulent times for markets and regulators, the consistent and dependable assessment of the individual and systemic risks of banks and sovereigns is of utmost importance. As the European Central Bank sets the focus of systemic risk management on the *macrofinancial* level (ECB 2010), one could conveniently observe the financial system as a portfolio of its member institutions, as noted by Lehar (2005). According to Gramlich and Oet (2011), within such a framework the role of a supervisor is twofold. First, to monitor the stability of the portfolio of banks conditional on their individual financial soundness, as well as the soundness of their interlinkages. Second, to analyze and assess the dynamics of the interconnectedness, especially in volatile periods, characterized by extreme events. The authors outline several conditions for a successful systemic risk model: *consistency*, *flexibility*, *forward-looking focus*, *correspondence with empirical data*, *suitability for the need of financial regulators*. This book introduces a comprehensive procedure to assess systemic risk of banks and sovereigns. We also introduce a number of intuitive measures of systemic risk to complement the existing toolkit of regulators.

With regard to policy decision-making, there are heated debates whether euro area sovereigns should be allowed to default. The proponents of this view claim that the economic costs of a sovereign default would be lower than the size of the bailout packages to keep the sovereign solvent. The systemic risk measures introduced in this book and especially $\Delta CoJPoD$ could be used to estimate the market expectations about the size of those alternative regulatory measures. To come up with meaningful regulatory suggestions for and against a bailout package, $\Delta CoJPoD$ should be coupled with an estimate of the economic losses to the euro area if the respective sovereign defaults. The resulting loss estimate should be used to determine the expected size of the bailout package. This expected size should be then compared to the welfare costs of alternative instruments in the regulatory toolkit. Note that even if

D. Radev, *Measuring Systemic Risk*, Studies in Systems, Decision and Control 409,
https://doi.org/10.1007/978-3-030-94281-6_6

a sovereign bailout package turns out to be optimal in order to minimize social costs, it might not be feasible even with the broadest possible international cooperation. Policymakers should then resort to their remaining tools to address the consequences of a sovereign default.

References

ECB: Financial stability review. June 2010. European Central Bank (2010)

Gramlich, D., Oet, M.: The structural fragility of financial systems: analysis and modeling implications for early warning systems. J. Risk Financ. **12**(4), 270–290 (2011)

Lehar, A.: Measuring systemic risk: a risk management approach. J. Bank. Financ. **29**(10), 2577–2603 (2005)